Synthesis Lectures on Games and Computational Intelligence

Series Editor

Daniel Ashlock, Guelph, ON, Canada

This series is an innovative resource consisting short books pertaining to digital games, including game playing and game solving algorithms; game design techniques; artificial and computational intelligence techniques for game design, play, and analysis; classical game theory in a digital environment, and automatic content generation for games.

Matthew Guzdial · Sam Snodgrass ·
Adam J. Summerville

Procedural Content Generation via Machine Learning

An Overview

Matthew Guzdial
University of Alberta
Edmonton, AB, Canada

Sam Snodgrass
Modl.ai
Copenhagen, Denmark

Adam J. Summerville
The Molasses Flood
Claremont, CA, USA

ISSN 2573-6485   ISSN 2573-6493   (electronic)
Synthesis Lectures on Games and Computational Intelligence
ISBN 978-3-031-16721-8   ISBN 978-3-031-16719-5   (eBook)
https://doi.org/10.1007/978-3-031-16719-5

This Springer imprint is published by the registered company Springer Nature Switzerland AG
The registered company address is: Gewerbestrasse 11, 6330 Cham, Switzerland

Preface

In October 2016 at the very specifically named Embassy Suites by Hilton San Francisco Airport—Waterfront in Burlingame, California the three authors of this book stood around a cocktail table. We were all Ph.D. students at the time, and we had all come to realize we were working on roughly the same thing. Some of us called it learning-based approaches to PCG, some called it machine learned PCG, and some called it automated game design knowledge acquisition (a name nearly as wordy as the hotel's). However, we all understood that we'd hit on something new and exciting.

All of us have graduated since that conversation in that hotel. But, we all still find this area exciting and are actively working to push it forward. We wrote this book to share that excitement with you.

This book was a collective effort, not just from the three equally contributing authors. This book could not exist without the ongoing and vibrant PCG and PCGML community, both in and outside of academia. Thank you all for your passion, your scholarship, and your enthusiasm.

Edmonton, Canada Matthew Guzdial
Copenhagen, Denmark Sam Snodgrass
Claremont, USA Adam J. Summerville
July 2022

Acknowledgments

Matthew would like to thank his fellow co-authors for agreeing to this immense undertaking, especially during the nightmare of the COVID pandemic. He'd also like to thank his husband, Jack, and the rest of their family for their endless enthusiasm and support. In particular, he'd like to thank his father, Mark, and his mother, Barbara, both rock star academics in Computer Science Education. You are a continual inspiration to him, and Matthew's glad to join you as published textbook authors. He would like to extend thanks to his students, colleagues, collaborators, and mentors. His life wouldn't be what it is without you all.

Sam would like to thank his wife, Jocelyn, for being incredibly supportive. From accommodating his strange hours when he was writing after work, to doing an enormous amount of labor to keep our day-to-day lives in order, and just generally making him happy in his life; Sam's contributions to this book would not have been possible without her. Thank you. He would also like to thank his family for their support and excitement around this project. And he would like to thank his colleagues, collaborators, and mentors who helped him get where he is. Of course he would also like to thank his cat, Sterling, for keeping him company during some weekend writing sessions. Lastly, he would like to thank his co-authors who, despite the time pressures and hellscapes, have been an absolute joy to write with.

Adam would like to thank his family for being so understanding and supportive during the journey to write this book. The throes of COVID have made finding the time, energy, and mental capacity to write this book extremely hard, and he knows this has not been easy for his children, Clark, Campbell, Maeve, and has been even harder for his wife, Mallorie, whose amazing work kept the family alive and afloat during the difficult authoring period. Mallorie, this book would not exist without you (or at least, Sam and Matthew would have had to write two more chapters each). He would also like to thank the mental health professionals who have helped him, specifically Laurie Ebbe Wheeler (he would also like to thank Vyvanse, without which editing would have been much harder). He would also like to thank his colleagues and mentors, including but not limited to, Michael Mateas, Noah Wardrip-Fruin, Ben Samuel, Joe Osborn, and James Ryan.

Finally, he would like to thank Matthew for subtweeting academic publishers and Sam for being a jovial, softening presence for his sometimes prickly co-authors.

We would all like to extend our heartfelt thanks to all reviewers and early readers of this book. We'd like to thank Dr. Jialin Liu and the other initial reviewer who chose to remain anonymous. In addition, we'd like to thank the 2022 members of Guzdial's GRAIL Lab for their detailed feedback and "book playtesting," particularly Adrian Gonzalez, Kristen Yu, Johor Jara Gonzalez, Anahita Doosti, Mrunal Sunil Jadhav, Akash Saravanan, Dagmar Lofts, Vardan Saini, Emily Halina, Kynan Sorochan, Jawdat Toume, Natalie Bombardieri, and Revanth Atmakuri. We'd especially like to acknowledge the use of game assets by Kenney.[1] Without his freely available game assets the figures in this book would look much worse.

The authors would like to dedicate this book to the memory of Dan Ashlock. Dan was a hugely influential figure in the academic games field. Dan believed in this book and went to bat for us with the publisher to get the book up and running. It is safe to say that this book would not have happened without his support, and we owe him a great thanks. Our field is diminished by his passing. We would also like to thank Joseph Brown for the introduction to Dan, and for serving as another instrumental force in making this book happen.

July 2022   Matthew Guzdial
                                                                                        Sam Snodgrass
                                                                                Adam J. Summerville

---

[1] https://www.kenney.nl/assets

Contents

About the Authors

Matthew Guzdial is an Assistant Professor in the Computing Science Department of the University of Alberta and a Canada CIFAR AI Chair at the Alberta Machine Intelligence Institute (Amii). His research focuses on the intersection of machine learning, creativity, and human-centered computing. He is a recipient of an Early Career Researcher Award from NSERC, a Unity Graduate Fellowship, and two best conference paper awards from the International Conference on Computational Creativity. His work has been featured in the BBC, WIRED, Popular Science, and Time.

Sam Snodgrass is an AI Researcher at Modl.ai, a game AI company focused on bringing state-of-the-art game AI research from academia to the games industry. His research focuses on making PCGML more accessible to non-ML experts. This work includes making PCGML systems more adaptable and self-reliant, reducing the authorial burden of creating training data through domain blending, and building tools that allow for easier interactions with the underlying PCGML systems and their outputs. Through his work at Modl.ai he has deployed several mixed-initiative PCGML tools into game studios to assist with level design and creation.

Adam J. Summerville is the lead AI Engineer for Procedural Content Generation at The Molasses Flood, a CD Projekt studio. Prior to this, he was an Assistant Professor at California State Polytechnic University, Pomona. His research focuses on the intersection of artificial intelligence in games with a high-level goal of enabling experiences that would not be possible without artificial intelligence. This research ranges from procedural generation of levels, social simulation for games, and the use of natural language processing for gameplay. His work has been shown at the SF MoMA, SlamDance, and won the audience choice award at IndieCade.

Introduction

1

This book focuses on **Procedural Content Generation via Machine Learning** (**PCGML**), the generation of media or content with machine learning techniques [195]. While machine learning (ML) has been used to generate a wide variety of content including visual art, music, and stories, PCGML (and this book) focus largely on video games. Therefore, we focus on the types of content specific to video games, such as levels, mechanics, game character art, sound effects, game narrative, and so forth. Video games are a difficult medium for ML, due to their complexity and lack of training data as we'll discuss further below. But this difficulty is part of what makes the problems in this field so compelling.

PCGML takes many different forms depending on the kind of content we wish to generate and the ML technique we apply. But for the purposes of an introduction you can imagine taking some amount of game content (levels, mechanics, etc.), training an ML model on this data, and then using the trained ML model to generate more content similar to the training data. This basic, intuitive process approximates many of the techniques discussed in this book. However, it also demonstrates some core problems with PCGML. How do we generate content that is not just a small variation on what we already have? How can we ensure the output has the desired characteristics of the input? What do we do if we don't have any data, or only a small amount? And so on.

In this chapter, we'll introduce the very basic concepts of PCGML in terms of its two component parts: Procedural Content Generation (PCG) and Machine Learning (ML), along with a discussion of the relationship between the two. We'll then discuss a brief history of PCGML as a means of situating this book in this constantly evolving area, and identify our intended audience for this book. We'll end by briefly outlining the rest of this book, and some various ways to read it. Readers with some expertise in PCGML or a related area may wish to start with the final section of this chapter.

© The Author(s), under exclusive license to Springer Nature Switzerland AG 2022

M. Guzdial et al., *Procedural Content Generation via Machine Learning*,

Synthesis Lectures on Games and Computational Intelligence,

https://doi.org/10.1007/978-3-031-16719-5_1

1.1   Procedural Content Generation

Procedural Content Generation (PCG) refers to the algorithmic generation of game content. By game content, we mean the various component parts of a game, including scripts or game code, levels or maps, character sprites or 3D models, animations, music, sound effects, and so on. By algorithmic generation we indicate some process or set of rules rather than typical human creation. Most often "algorithmic" indicates the use of computer code, and that's the way we'll discuss it in this book, but it could involve any set process or rules, such as generation via cards or dice [55]. PCG can look like the landscapes of Minecraft, the conversations and dungeons of Hades, or almost everything present in No Man's Sky.

PCG in the video game industry is most commonly used during development time, when it is used at all. That means that most PCG is invisible to the players of games. For example, most modern, open-world games make use of some amount of PCG to create their worlds. For example, the developers of the often-re-released Skyrim generated the game world's landscape with PCG. From there, human developers went back through the landscape, tweaked it, and added extra content like decorations, 3D models, enemies, quests, and so on. PCG is sometimes visible to players, particularly when it shows up inside games "at runtime," as in the examples of Minecraft, Hades, and No Man's Sky, though this is less common.

The examples we have given thus far are examples of classical PCG. Classical PCG, which we will discuss further in Chap. 2, relies on classical Artificial Intelligence (AI) methods like grammars and search [160]. Classical PCG has seen some adoption by the video game industry, and even in other, related industries. For example, Speedtree, a library that allows designers to quickly generate many unique 3D tree models with PCG, has been used in many AAA games, and even in Hollywood films like the Avengers series. However, most game content is made without PCG. In fact, PCG sometimes sinks projects, such as in the case of Mass Effect: Andromeda, a follow-up to the Mass Effect franchise which lost nearly two years of development time to an attempt to create a procedural galaxy in which the game would take place. The issue here is that working with PCG is tricky, it takes specialized design and algorithmic knowledge, along with buy-in from the entire development team. Therefore, it can often take more time and resources than just developing the game using more standard industry practices.

This design and algorithmic knowledge is the core requirement in creating a PCG generator. Specifically, PCG requires that the user hand-author knowledge to construct their generator so that it can output the kinds of content the user wants, and not the kinds they don't. In a grammar, a developer might need to author chunks of content and rules for how they fit together. You can think of this like making individual Lego bricks, which can then be used to create many different structures. In a search-based approach, a developer instead needs to author a representation of the search space (a space where every point is a piece of content), neighbor functions defining how to move through that space, and a fitness function to indicate what high quality content looks like in that space. More on both of these types of approaches in Chap. 2. Tweaking this knowledge is key to shaping the output of the algorithm, and is an art in and of itself. As Kate Compton famously put it, it's difficult to solve the "10,000 bowls of oatmeal problem" [26]. It's easy to generate a lot of *something*, but it's tricky to make that something *interesting*.

Intuitively, we might consider learning this knowledge instead of having to hand-author it. After all, there's a large number of existing, high quality games. If we could learn to design based on the content from these existing games, we might be able to empower more people to benefit from PCG. This would allow more people to make games, and even for the creation of new types of games and experiences that would be impossible or impractical with modern game development practices. For example, consider how modern "open world" games are still limited to a single region, how games could adapt to their players, or even how players could create their own content for the game as they play. The list goes on, and we'll discuss the future potential for PCGML further in Chap. 12.

1.2   Machine Learning

Machine Learning (ML) refers to algorithms that "learn" the values of variables from data or experience. While it's an AI approach that can lead to amazing things, there has been a great deal of misinformation spread about machine learning. Essentially, it is just a way of adapting a function based on data. As a simple example, let's say we have the function $w * x + b = y$. This function takes in an x as an input argument, which should be a number, and outputs another number y. There are two variables in this function: w and b. Depending on the values of w and b we'll get different outputs for different x inputs. If you haven't recognized it already, this is just the function to describe a line. If we have enough examples of x inputs and their associated y outputs, we can approximate what the best values of w and b would be to match these. We're approximating a function ($w * x + b = y$) to match some training data (our pairs of x's and y's). You can see a visualization of this, with slightly different variable names, in Fig. 3.4. While there's lots of different kinds of machine learning, and at times the functions can get pretty complicated (to the point where we forget they're just functions and we start calling them models), these same basic principles stand. We'll focus on machine learning from Chap. 3 onward.

Modern ML approaches tend to struggle on a number of types of problems: (1) problems with low amounts of training data, (2) problems where the data has high **variance** (a lot of differences between pieces of data), and (3) problems without clear metrics for success. PCGML includes all three of these types of problems. There is typically a very small amount of training data available for a particular kind of game content, in comparison to datasets of non-game content. ImageNet, a common image classification dataset for machine learning, contains roughly 1.3 million images. In comparison, there are estimated to only be about 1.2 million published video games available for purchase [185]. Compared to images, video games are much more complicated and differ much more from one another. Because of that, we'd actually expect to need significantly more data to model games compared to images. Further, across these 1.2 million published video games there's no consensus about what makes a good game. This is a positive thing, as different games are better suited to different people. However, this does mean there are no clear metrics for things like how fun a particular type of game content might be. This means that, in most cases, we can't just optimize for game quality.

The problems that PCGML confronts us with aren't just limited to games. In fact, anytime we want to model the output of individual humans these problems arise, since individuals can only produce so much data, humans differ from one another, and there's limited metrics for replicating human evaluation. Thus, solving these problems for PCGML can allow us to push the boundaries of what is possible with ML broadly.

1.3   History of PCGML

Procedural Content Generation via Machine Learning (PCGML) is the algorithmic generation of game content using machine learning methods. It was proposed to try to address problems in PCG and ML, and has enjoyed a great deal of popularity since. However, it's still a very young field, and we'll try to reflect that in this section and throughout this book.

In 2013, Sam Snodgrass and Santiago Ontañón published "Generating Maps using Markov Chains" the first paper later recognized as an example of PCGML [176]. In it, they discussed a project training a Markov Chain (a type of probability-based ML model we'll discuss more in Chap. 6) on Super Mario Bros. levels in order to generate new Mario levels. In this paper, the authors simply referred to their approach as a "learning-based approach to PCG." Other level generators had used machine learning as part of the generation process, such as the level generator of Robin Baumgarten from the 2010 Mario AI Championship Level Generation track, but this generator still relied on hand-authored chunks of levels, and sequenced these hand-authored chunks based on an ML analysis of player behavior [157]. Similarly, the Ludi system took in existing game content (in this case whole board games) as input in order to generate new board games, but no learning occurred [20]. Instead, the system recombined the existing games without altering anything about the generation approach based on the input. That's why we point to Sam and Santi's 2013 paper as the beginning of PCGML.

From 2014 to 2016, a large number of early PCGML systems debuted. A number of additional Markov Chains methods were published [33, 180]. Researchers began to focus on the problem of acquiring sufficient training data [59, 200], and the first neural network PCG experiments were published [77, 198]. At this point, the authors of this book and a large cadre of other early PCGML researchers began work on a survey paper of this growing area. Together, we would dub it Procedural Content Generation via Machine Learning [195].

Since the survey paper, PCGML has continued to grow as a research area. WaveFunction-Collapse, a simple and low-data PCGML approach, began to gain popularity as an approach among indie game developers before being picked up by academic researchers [88]. There were the first attempts at generating entirely novel games with PCGML [63, 150], and at trying to create PCGML tools for designers [35, 57, 154]. But overall, the fundamental problems of PCGML remain unsolved, including the problems discussed above and many more remaining. We will discuss some of these open problems in Chap. 12.

1.4   Who is this Book For?

Our hope is that this book is accessible to PCG practitioners, ML practitioners, and anyone interested in these topics. The book can be used as the basis for a class, with every chapter serving as the basis of 1-2 lectures, as an introduction to these topics, or simply as a reference or guide. Our hope is that this book can demystify ML for those on the game design and PCG side of things, and make the benefits of applying ML to PCG clear for those on the ML side of things. For a class, we have written this book to be programming language agnostic, but it will require at least some understanding of coding. We recommend using this book for students at least at the undergraduate level, as many of the concepts in the book rely on fairly complex mathematics, though we'll do our best to express these clearly. Our hope is that the individual chapters can serve as a reference and guide for individuals looking to implement particular PCGML approaches, or for those interested in conducting PCGML research.

1.5   Who is this Book *Not* For?

While it may seem natural to some readers, we won't be covering reinforcement learning for automated game playing (the technology behind AlphaGo, OpenAI Five, AlphaStar, etc.) in this book. We will cover how reinforcement learning can be applied to PCG in Chap. 10, but not how to train agents to play or interact with existing content. There are many excellent resources on this subject, but this is not one of them. We focus on game design problems, not game playing problems.

We also do not intend this book to be an all encompassing look at Procedural Content Generation or Machine Learning as individual fields. We instead focus on the intersection of these two fields. While we'll introduce concepts from both as they are relevant to PCGML, if you find you want to dig deeper we recommend seeking out introductory texts on PCG [160] and/or ML [125].

1.6   Book Outline

In this section we'll briefly discuss the chapters of the book and some suggested reading orders depending on your level of familiarity. If you are already familiar with PCG, you can safely skip Chap. 2. Similarly, if you are already familiar with ML, you can safely skip Chap. 3. We recommend reading Chap. 4 regardless of your familiarity with these topics, as we overview the PCGML project process we'll use in this book. From there, readers can skip around as they like depending on their level of familiarity with the chapter topics. However, newcomers would likely benefit from reading the chapters in order. We recommend ending with Chaps. 12 and 13, regardless of your reading order. The chapters will cover the following topics:

- **Chapter** 2 presents an overview of "classic" approaches to PCG, which do not make use of machine learning.
- **Chapter** 3 covers the basic concepts necessary to understand the machine learning aspects of this book.
- **Chapter** 4 overviews our process for PCGML projects, along with covering practical and ethical considerations.
- **Chapter** 5 focuses on our first PCGML area: ML constraint-based approaches. This chapter covers the most commonly applied PCGML approaches in industry at the time of writing.
- **Chapter** 6 covers our second PCGML area: probabilistic models. These are some of the simpler PCGML approaches, particularly for those with a background in probability.
- **Chapter** 7 begins our coverage of deep neural networks (DNNs) for PCGML, starting with the basics. We recommend reading this chapter before Chaps. 8 and/or 9.
- **Chapter** 8 covers DNN models for processing sequences like text that can be applied to PCGML.
- **Chapter** 9 covers DNN models for processing image-like data structures that can be applied to PCGML.
- **Chapter** 10 focuses on our last major PCGML area: PCG via Reinforcement Learning or PCGRL. This differs significantly from the rest of this book due to not relying on existing training data.
- **Chapter** 11 introduces mixed-initiative PCGML, incorporating PCGML into tools for designers. This is an open problem but with existing applications, which we cover in this chapter.
- **Chapter** 12 overviews other open problems in PCGML (at the time of writing). These topics could serve as the basis for a research project or thesis.
- **Chapter** 13 ends with our conclusions, some discussion, and a variety of resources for PCGML practitioners.

Classical PCG

2

The other chapters in this book cover a wide range of approaches to procedural content generation that leverage different machine learning paradigms. Before jumping into the machine learning-based approaches, we will use this chapter to give a brief introduction to classical (i.e., non-machine learning-based) PCG approaches to provide context for the remainder of the book. In particular, we will introduce and discuss **constructive**, **constraint-based**, and **search-based** PCG as PCG paradigms that do not rely on machine learning. Constructive PCG relies on hand-authored rules and functions for assembling new pieces of content. Constraint-based PCG approaches define what a "valid" piece of content is using constraints, and use those constraints to find new content. Finally, search-based PCG defines the space of content, and uses optimization procedures to find high quality content within that space.

Each of the groups outlined above use unique methods for generating content. For each of these groups we will highlight the input needed from the user or developer, how that approach works at a high level, and examples of how related approaches have been or can be used. Lastly, we will discuss possible connections and extensions to PCGML. However, before we begin discussing these paradigms, we will give a brief overview of types of content we might want to generate.

2.1   What is Content?

When hearing about procedural content generation, you may think "What do they mean by content?" or "What can we generate?" The idealistic answer is that pretty much any part of a game (be it structural or mechanical) can be considered **content** or something that we could try to generate. Everything from game levels to textures to stories to gameplay mechanics to full games have been procedurally generated. A full categorization of content types is

M. Guzdial et al., *Procedural Content Generation via Machine Learning*,
Synthesis Lectures on Games and Computational Intelligence,
https://doi.org/10.1007/978-3-031-16719-5_2

outside of the scope of this book, but there are existing discussions around types of content. Hendrikx et al. [72] categorize game content hierarchically, starting at the bottom with *game bits* (i.e., the atomic game elements), then up to *game spaces* (i.e., the environment or world where the player and agents interact with the game), all the way up to *game designs* and through to *derived content* (i.e., content or information derived from the game, such as leaderboards). The detailed description provided by Hendrikx et al. makes this paper a good resource if you want a deeper discussion of content types. In the original PCGML survey paper [195] we (along with the other authors) instead focused on the representation of the content. We grouped content (regardless of its function) according to its structure: *sequence*, *grid*, or *graph*. This categorization could be useful when considering how to represent your chosen content type, and what implications that might have on the appropriate approach. Liu et al. [116] use a flat structure of content types: *game levels*, *text*, *character models*, *textures*, and *music and sound*. They give an overview of how different machine learning and especially deep learning methods have been applied across these categories of content; as such, it is a good resource if trying to decide on an approach or model architecture to use for a certain type of content. Notice, however, that the same content can be represented in many different ways, each of which lends itself to certain ML techniques. Similarly, disparate types of content might be represented in the same way, leading to similar ML applications. For instance, a level might be represented as an image (like textures), it might be represented as a sequence (like text), or it might be represented as a collection of content oriented in space (like a character model). As such, in this book we will tend towards representational categories (e.g., sequences, grids, graphs).

Categorizing content types can be a useful tool or lens through which to view PCG. But more importantly, while reading this book try to keep an imaginative mind. When we introduce a new approach we will give examples of how it has been used, and perhaps how it could be used in the future, in order to provide context and hopefully deeper understanding of the technique. But as you read try to also think of new ways the approach could be leveraged (e.g., new content types, new representations, new applications). PCGML is a young field, and there is a lot of unexplored space; so keeping an inquisitive eye open as you become acquainted with the field could lead to the next big innovation!

2.2   Constructive Approaches

Constructive procedural content generation describes the family of approaches that quickly generate a piece of content using rules and randomness, often in a one-shot fashion [158, 209]. Constructive approaches rely on the encoded design and domain knowledge from the creator of the approach. The approach then directly uses this encoded knowledge to create new content. Since these approaches directly rely on the encoded domain knowledge of the user, they allow the creator a lot of control over the generative process. Additionally, these approaches tend to be somewhat simple (in the algorithmic and conceptual space), making them more accessible and as such the most common family of approaches used in games. An example of a conceptually simple approach, is one where the designer creates two segments for a level, and a rule that says "flip a coin to decide which section should be placed next." The encoded knowledge in this situation is the hand-authored level segments as well as the rule for how to place them. Now, this toy example serves two purposes: to give some insight into how a simple constructive PCG system might work, and to highlight what is needed from a human designer of such a system. An important concept to remember here is that constructive PCG systems are often one-shot generators (i.e., will create a piece of content without relying on feedback or being expected to further improve on that generated piece). This has the implication that either (1) the encoded knowledge, rules, etc. need to ensure that "bad" content cannot be created, (2) the generator needs to be used in scenarios where low quality content is permissible, or (3) there is a human at the end of the system choosing the "good" outputs.

The toy example above is not representative of all constructive PCG and the encoded knowledge used in constructive PCG can take many forms. In the following subsections we will discuss constructive approaches that leverage an increasing amount of hand-authored encoded knowledge starting with **noise** (structured randomness), moving to **rules** (directly encoded knowledge), and closing with **grammars** (more formally structured rule encodings).

2.2.1 Noise

Noise refers to a family of structured random functions. In graphics and PCG noise functions can be described as functions that give random real values[1] in an interval (commonly, 0 and 1) over some domain (i.e., 1-dimensional, 2-dimensional, etc.). White noise is one of the most commonly known types of noise functions, and can be thought of as sequences of independent random variables over an interval (e.g., a grid with each value randomly chosen between 0 and 1) with constant density across all frequencies. In addition to white noise, there are other types of noise with various properties. For example, while white noise is uniformly distributed over frequency ranges (i.e., will have quickly changing values and slowly changing values with equal likelihood), pink noise has a denser distribution around the lower frequency ranges (i.e., slowly changing values are more likely than quickly changing values). Alternatively, Blue noise has a denser distribution around higher frequency ranges (i.e., quickly changing values are more likely than slowly changing values). If thinking of images, pink noise will tend to have smoother transitions between colors, such as going from a dark to light grey spread over many pixels; blue noise will tend to have transitions from light and dark more quickly in only 1 or a few pixels; and white noise will have both

---

[1] Here we mean *real* in the mathematical sense (i.e., a value in $\mathbb{R}$). That is, essentially a continuous number that can be written with infinite decimal point precision.

of these patterns occurring. A detailed survey by Lagae et al. [101] covers different types of noise as used in image processing and graphics.

In the context of constructive PCG, the knowledge being encoded by the designer when using a noise-based approach, is which noise function has the properties most useful for the domain and what the sampled noise represents in that domain. Blue noise, because of how it is sampled and distributed, can be useful for pseudo-randomly distributing objects somewhat evenly throughout a space or level [2], whereas pink noise might be better suited to generating landscapes with more slowly changing height values. Below we give a brief introduction to a few noise-based approaches to texture and terrain modeling, but the reader is referred to the book by Ebert et al. [41] for a more detailed look at these topics.

Texture and Terrain Synthesis

Noise functions are commonly used in the creation of procedural textures and terrain (or heightmaps). Specifically, generating natural textures such as wood, marble, and clouds often rely on the randomness and detail provided by noise functions [101]. Examples of this can be found as far back as the 1980s in the graphics community where applying noise to textures led to more realistic looking natural textures [136]. Since then, there has been much more work in procedural textures and image synthesis using noise functions [6, 80, 102], as well as in terrain generation [5, 168].

A common extension to noise-based approaches leverages a hierarchical (or multi-resolution) strategy. In these multi-resolution approaches, noise is sampled at various resolutions and combined together to form the final result. For example, we can use a white noise function to sample values between 0 and 1 on a 16 × 16 grid (Fig. 2.1 first). Next, we use the same noise function to sample values between 0 and 0.5 on a larger 32 × 32 grid (Fig. 2.1 second), 0 and 0.25 on a larger 64 × 64 grid (Fig. 2.1 third), and finally 0 and 0.125 on the largest 128 × 128 grid (Fig. 2.1 fourth). In each of these steps, we double the height and width of the grid and halve the range of the noise values; this forces the approach to create different scales of features (i.e., bigger features in the initial grid, and smaller details in the

Fig. 2.1 This figure shows sampled noise at different resolutions and magnitude ranges. The left-most grid was sampled at a 16 × 16 resolution with a range of 0 to 1, with the values between filled in with bicubic interpolation. The second grid was sampled at a resolution of 32 × 32 with a range of 0 to 0.5, and interpolated in the same way. The third grid was sampled at a 64 × 64 resolution with a range of 0 to 0.25, and the fourth grid was sampled at a resolution of 128 × 128 with a range of 0 to 0.125. The final grid is the result of averaging the grids together, which could then be used for a terrain heightmap or a texture

larger grids). We then scale up the lower resolution grids (i.e., the 16×16, the 32×32, and 64×64 grids) to the full size (i.e., the 128×128) by interpolating the values between the sampled points. Lastly, we combine the values at each position together, which gives us a texture or terrain with features at different levels of granularity (Fig. 2.1 last). In our case, we averaged the values together, but adding and normalizing is also common. This final grid can be used as a heightmap for terrain where brighter pixels are higher altitudes and the darker pixels are lower altitudes. We could even translate the darkest pixels to water or being below sea level.

2.2.2 Rules

Rule-based PCG techniques are a subset of constructive approaches that create content by leveraging a set of manually-defined rules, and often rely on either designer-created chunks of content or designer-created templates. Rule-based PCG systems are among the most commonly used PCG approaches in commercial games due to the high amount of control and designer input they are able leverage.

For example, we can imagine a scenario where a designer has created a set of different rooms that a player might interact with and investigate. Some of these rooms might require the player's character to fight enemies, some might have treasure chests, some might allow the player's character to purchase items, and some might have special events that get triggered when the player enters. The designer doesn't want the players to experience the same sequence of rooms every time, and so instead of placing the rooms in a set layout, they devise some rules for how the rooms can be laid out (e.g., only place 1 treasure room in the map, don't place more than 5 combat rooms near each other, don't place extra strong enemies too early in the map, only allow certain types of events under different scenarios, etc.) This rule-based constructive PCG approach relies on the designer encoding their knowledge and desires for the game into smaller content blocks (rooms) as well as into the rules for how to place the content blocks. This is a fairly common approach in rule-based PCG approaches, and in fact, the deckbuilding game *Slay the Spire*, and the dungeon crawling game *Hades* follow approaches similar to this for generating their maps.

As another example, in *No Man's Sky* each of the planets in the universe are procedurally generated along with their associated biomes and lifeforms. *No Man's Sky* uses a "blueprint" system, where hundreds of basic templates for animals, plants, etc. are first defined by artists and designers. Then during the generation of a planet, the biomes and environment are first created using a set of rules. The planet and its biomes are then populated with instantiations of the base templates for animals and plants. The instantiations are also made using a set of rules to ensure consistency across creatures/biomes/plants/color palettes/etc. In this case the designers and artists encode their knowledge into various templates, content blocks, and systems of rules, but encoding rules and creating templates can be very difficult to get right. This can be seen with the improvement of *No Man's Sky* post-release, partially due to the designers tuning the rules and templates.

While the examples above present rule-based approaches for world and creature generation, rule-based approaches can be applied to any type of content, provided the relevant knowledge can be encoded into the rules. *Skyrim* uses a rule-based approach for generating what they call "radiant quests." These quests rely on a templates that could be of the form "go to ⟨location⟩ to kill ⟨NPC⟩ and get ⟨item⟩ as a reward." Those locations, enemies, rewards, etc., are then randomly set based on the player's current level and previous actions. Rule-based approaches are broad enough to cover essentially any content type. However, sometimes a more structured approach can be useful for representing the rules and content; and in these cases we can use grammar-based approaches.

2.2.3 Grammars

Grammar-based PCG methods can be thought of as a subset of rule-based approaches since grammar-based approaches rely on the application of a set of rules to create content. However, where general rule-based approaches can be formatted and defined in arbitrary ways, grammars are defined according to specific formalizations. These formalizations can be useful for providing a framework for a designer to encode their knowledge. A **generative grammar** consists of a set of **terminal** (can't be changed) and **non-terminal** (must be changed) **symbols**, and rules describing how those symbols can be produced. A symbol in this case is just a way of representing certain information about the game or content. For example, if we are trying to generate platforming levels, then we can have terminal symbols (e.g., characters or objects) representing different hand-authored sections of levels; or if we are generating desired player paths through a dungeon then we can have symbols representing the player turning left, turning right, and going straight.

To demonstrate how a generative grammar can work, we'll walk through a simple grammar for generating text strings. This grammar will consist of a single non-terminal symbol, A; two terminal symbols, b and c; and the following rules:

1. $A \rightarrow bAc$
2. $A \rightarrow bc$

Using this grammar, if we start with the string A, then we can apply the first rule to replace the A with bAc. We can further apply the second rule to replace the A in the resulting string with bc, to get the string $bbcc$. With this resulting string we are unable to apply any of the rules, and so $bbcc$ is the output of applying this sequence of rules. If we look at the rules and work through some examples we can see that this grammar can only produce strings consisting of some positive number of b's followed by the same number of c's. The above grammar is a toy example, but grammars can be used to create more complex content, such as plants [138], missions [37], and dungeons [205] to name a few.

Grammars can model more complex structures by adding more varied rules, changing the symbols used by the grammar, and changing what the symbols represent. For example, using the same grammar above we can have the terminal symbols b and c represent a hand-crafted section of platforming level. Then, using that grammar to generate a string is equivalent to generating a new platforming level. Furthermore, we can add more symbols to represent additional hand-crafted level sections, d and e. We can also add more rules to increase the **expressivity** (i.e., the size of the generative space) of the grammar:

3. $A \rightarrow Ad$
4. $A \rightarrow ed$
5. $A \rightarrow eeA$

which allows us to produce strings such as $A \rightarrow eeA \rightarrow eebAc \rightarrow eebAdc \rightarrow eebeddc$.

The grammars discussed so far have all used characters for the symbols and generated strings. However, grammars can be used to represent and generate many different types of content and are not limited to generating strings. One such example is using **graph grammars** to generate missions for a game [37]. In this work, the symbols of the grammar are graph nodes representing possible steps for a mission, and the rules define how that graph can be grown by adding new nodes and connections.

Within rule and grammar-based constructive PCG, there is room for PCGML adaptations. For example, we can try to extract the rules for how content blocks are placed together from existing examples. Or we can try to learn what the size and structure of content blocks for a given domain should be. In later chapters we will discuss how we might try to learn such rules and structures.

2.3   Constraint-Based Approaches

Constraint satisfaction for procedural content generation describes a group of techniques that represent the generative space as a set of variables with potential values, and a set of constraints (or facts that must be true in the final generated piece of content). In the context of constructive PCG, these approaches require the designer to carefully encode what they want or believe to be true about the domain (i.e., the constraints), and a representation for the domain (i.e., the variables and what they represent). Constraint-based PCG approaches have previously been used for generating platforming levels [173, 174], puzzles [169, 170], and terrain [167].

As a concrete example, let's define a constraint satisfaction approach for generating palindromic strings (i.e., strings that are the same frontwards and backwards, such as, *dad* or *racecar*). We can define this approach with the following setup:

- **Variables**: $char_0$, $char_1$, ..., $char_n$
- **Possible Values**: a, b, c, d, e, f, g, h, i, j, k, l, m, n, o, p, q, r, s, t, u, v, w, x, y, z
- **Constraint 1**: $char_t = char_{n-t}$
- **Constraint 2**: [$char_0 char_1 ... char_n$] is a valid word

If we start with 5 variables, then one run of the generator could work as follows:

1. Start with $char_0$, $char_1$, $char_2$, $char_3$, $char_4$
2. Choose $char_1$ at random (since all have the same set of possible values presently).
3. Set the value of $char_1$ randomly according to its possible values. $char_1 = $ "a"
4. Propagate the effects of setting this variable.

   - **Constraint 1**: We know that $char_1 = char_3$.
   - **Constraint 2**: We could check a dictionary and track all 5 letter words with "a" in the second and fourth positions. This could give us the list: "banal," "basal," "cabal," "cacao," "canal," "carat", "fatal," "kayak," "macaw," "madam," "nasal," "naval," "pagan," "papal," "radar," "rajah," and "salad." This list reduces the possible letters for each position.
     - $char_0$ can be "b," "c," "f," "k," "m," "n," "p," "r," or "s."
     - $char_2$ can be "b," "c," "d," "g," "j," "l," "n," "p," "s," "t," "v," or "y."
     - $char_4$ can be "d," "h," "k," "l," "m," "n," "o," "r," "t," or "w."
   - Here we could also do a another pass of **Constraint 1** to further reduce the possible values for $char_0$ and $char_4$ by removing any letters that are not present for both variables. But for simplicity we will skip this.

5. We now have $char_0$ a $char_2$ a $char_4$
6. Choose another position. This time we will choose the variable with fewest remaining values: $char_0$. Recall, $char_0$ had its set of possible values reduced to 9 letters in the previous step. Note that this is just one possible selection approach.
7. Set the chosen variable's value from the set of possible characters for $char_0$, and get $char_0 = $ "r"
8. Propagate the constraints again:

   - **Constraint 1**: We know that $char_4 = char_0$. And so "r" is the only possible value for $char_4$.
   - **Constraint 2**: We check the dictionary and track all 5 letter words with "a" in the second and fourth positions and "r" in the first and last. This gives us the list: "radar," which reduces the possible letters for each position. Note that many solvers will stop here and return "radar" since it is only valid solution left, but we will continue through the steps.

     - $char_2$ can be "d"

9. We now have ra $char_2$ ar
10. Only one variable left, $char_2$
11. Set $char_2$ according to constraints, $char_2$="d"
12. We now have "radar"
13. No more variables left to set, so we can now check **Constraint 2**.
14. "radar" is a valid word, so we are done.

The above example was meant to show how a simple constraint satisfaction procedure could function, but in practice, many constraint satisfaction approaches follow a similar procedure:

1: Initialize a set of variables for a possible solution
2: **while** at least one variable is not set **do**
3:     Select which variable to set next
4:     **if** selected variable has 0 possible values **then**
5:         Generation failed
6:     **else**
7:         Set selected variable's value from possible values
8:         Update all other unset variables' possible values
9:     **end if**
10: **end while**
11: Return solution

In line 3 in the pseudo-code, we do not specify *how* to select which variable to set a value for next. The reason for this is that there are in fact several different approaches for choosing the next variable, such as choosing the most or least constrained variables, or choosing randomly. Similarly, in line 7 we also do not specify *how* to select the value for the variable. This is, again, because there are different methods for choosing the value when there are multiple possibilities, such as choosing the least constraining value. These different methods each have their benefits and drawbacks, and so it is worth considering which approach to use when working with constraint-based PCG. However, discussing these differences in detail is beyond the scope of this chapter. Karth and Smith [90] explore the impact of some of these different settings on a specific constraint-based PCGML approach.

Using a constraint satisfaction approach for a PCG problem has the benefit of allowing the designer to explicitly state what should be allowed or disallowed in the content. But as with using grammar-based approaches, these approaches require a specific formulation of the design space. As an example of how constraints might affect output, we can look at how Smith and Mateas [170] used answer set programming (ASP) to generate chromatic mazes. Chromatic mazes are grids with their cells filled with different colors, and where traversal between cells depends on the colors' relations in a color wheel (e.g., the player can only move to an adjacent cell if the colors of the starting and ending cell are adjacent in the provided color wheel). In this work they define general constraints for the domain, such as, each cell must be filled with a color, the allowed colors and their relations, and that there must be a starting and ending cell. They define further constraints to ensure that each generated maze contains a valid path from the start to end cell. Just these constraints are enough already to generate valid chromatic mazes. However, there are no guarantees about the difficulty of the mazes or length of the path needed to solve the maze. For additional control here, they add another constraint that specifies a minimum length for solution path. This additional constraint can greatly reduce the set of possible mazes, but does so in a way that guides the generator towards more desirable output.

Choosing a good representation and reasonable constraints can be difficult; it depends greatly on the domain and the desired features of the content. One way to address the difficulty of devising constraints for a domain is to learn them from data. In Chap. 3 we discuss how machine learning can be used to support constraint satisfaction approaches and vice versa.

2.4   Search-Based Approaches

Search-based procedural content generation (SBPCG) [209] describes a category of PCG approaches that use search algorithms to generate content. In SBPCG the designer is expected to encode their knowledge to define what the space of possible content looks like, and how to find a "good" piece of content in that space. More formally, the designer encodes their knowledge into:

- the **search space**, which defines the possible content to be generated, and
- the **search procedure** which defines how a piece of content or solution will be selected from the search space.

Oftentimes the search space leverages an abstracted representation of the desired content, which allows for a more efficient search of the space. The representation chosen here has a large impact on the type of content that can be created.

For example, let's say we are trying to generate weapons for a game. A simple encoding into a search space could use two variables to represent a weapon: damage, and time to attack; both of which are values between 1 and 100. This abstract representation reduces the space of all possible weapons into a two-dimensional space representing damage per second (DPS). However, going back to the importance of choosing an appropriate representation, if we are interested in the size or type of the weapon this representation gives us no mechanism for exploring that feature.

For now let's assume we are only interested in the DPS of the generated weapons, and thus the representation we have chosen is perfect. We now need to define how we want to search the space for a solution. The search procedure can use any search algorithm (e.g., exhaustive

Table 2.1 This table shows 5 example iterations of a hill climbing algorithm using our representation and fitness function, where we are trying to maximize the fitness score. The current values and fitness at each iteration are shown in the first 3 columns. The final four columns shows the potential steps that can be taken by the algorithm and their resulting fitnesses

Damage	Time	Current fit	−1 Damage	+1 Damage	−1 Time	+1 Time
25	60	110	108	**112**	109	111
26	60	112	110	**114**	111	113
27	60	114	112	**116**	113	115
28	60	116	114	**118**	115	117
29	60	118	116	**120**	117	119

search, heuristic search, stochastic optimization, etc.) to explore the search space. However in practice, evolutionary algorithms (EAs) are most commonly used in SBPCG [43, 85, 207]. EAs, hill climbing, and stochastic optimization as search procedures already align well with exploring an abstracted search space, and they allow for optimizing particular features. In the following section we will describe evolutionary algorithms as they apply to PCG, so in our current example we will show how we can use a hill climbing approach.

In hill climbing algorithms, we start with a random solution and at each step the algorithm makes an incremental change that tries to improve the solution. Let's say we want to generate a weapon that is slow, but powerful. Then we can define a **fitness function** (i.e., an evaluation function used for guiding the search), as:

$$fitness(damage,\ time\ to\ attack) = 2 \times damage + time\ to\ attack,$$

which rewards a solution if it deals a lot of damage and if it takes a long time to attack. Table 2.1 shows a random initialization for a weapon, and several iterations of improvement using a greedy hill-climbing approach that just selects the change that gives the biggest improvement in the fitness function.

Notice that with this fitness function and search space, the algorithm can just continue until it finds the best possible solution (e.g., 100 damage and 100 time to attack), or the **global optimum**. However, the algorithm is only guaranteed to be able to find the global optimum because there are no **local optima** for it to get stuck in. A local optima is the best solution given all its neighbors (i.e., the search procedure can't find a better solution nearby). In hill climbing, this means a solution in the search space the hill climber is unable to improve upon with the small changes it is able to make, but which is not representative of the best possible solution. For example, if instead of our current fitness function we use something more complex, like:

$$fitness(damage,\ time\ to\ attack) = damage^3 - damage + time\ to\ attack,$$

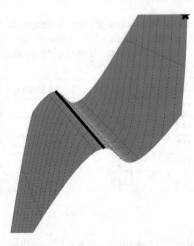

Fig. 2.2 This figure shows a plot of the previously discussed fitness function: *fitness(damage, time to attack) = damage^3 − damage + time to attack*. Notice the local maximum near the center of the plot (black line) and the global maxima in the top-right (black star)

then the resulting function landscape has some bumps in it, which we refer to as a local optima. Figure 2.2 shows this function. In this scenario, if we use a greedy hill climbing approach, then the initialization of the values can impact the final output. If our initial values start us closer to the small hill than to the global optima, then our greedy hill climbing approach will get us to the local maximum at the top of that small hill, but miss out on the global maximum. Local and global optima, and generally optimization, are also important for a different type of search algorithm that is often used for PCG: evolution.

2.4.1   Evolutionary PCG

Evolutionary algorithms (also called genetic algorithms) have been explored extensively in search-based PCG research [209] and used to generate content for several games (e.g., *Petalz* [145], *Galactic Arms Race* [69]). In general, evolutionary algorithms (EAs), are define by their

- **phenotype**: the abstracted representation of a piece of content that defines the search space and which the algorithm uses during search and evolution
- **genotype**: the final piece of content that is represented by the phenotype
- **fitness function**: an evaluation function that guides the search.

Continuing with the biological evolution metaphor further, we can intuitively liken the **phenotype** to the DNA of a biological organism (e.g., a gene for hair color), and the **genotype** to organism itself (e.g., does it have dark or light hair?). And tying back into our previous definitions, the phenotype is equivalent to the search space. Thus, as with other search-based approaches, the knowledge of the designer is encoded into the representation of the search space, and the fitness function that guides the search. As mentioned, the interesting difference between the search approaches discussed above (i.e., hill climbing, and optimization) and Evolutionary Algorithms (EAs), is the method in which the search is performed. Recall that the hill climbing approach discussed earlier ran into issues when there were local optima in the search space. EAs alleviate this problem by enabling bigger steps through the search space. That is, the methods that EAs use to explore the search space are able to reach more distant neighbors at each step than a standard hill climbing approach is able to. These bigger steps are achieved using **crossover** and **mutation** operators on elements in the search space (i.e., phenotypes). The crossover operator typically involves two (or more) elements, and can reach distant neighbors by recombining parts of these phenotypes. The mutation operator typically involves one element, and reaches nearby neighbors by making a small change to the current phenotype.

Now that you have a sense of how EAs can explore a search space, let's walk through the approach in more detail. EAs start with a group of phenotype encoded solutions called the initial population (line 1 in the algorithm below). Each of these solutions is evaluated using the fitness function (line 2). Then, new members are added to the population by performing different varying operators on the existing population, such as combining multiple individuals (i.e., crossover) or changing an individual (i.e., mutation, line 4). The details of selecting which individuals to modify, and how exactly to modify them varies by algorithm and domain, but often the selection of individuals is based on their fitness in some way. After the new members are added, they are evaluated with the fitness function (line 5), and some subset of the total population is retained (line 6). The process of varying, evaluating, and retaining continues until a suitable solution is found (line 3).

1: Initialize a *population* of individuals in the phenotype representation
2: Evaluate the individuals in *population* using the fitness function
3: **while not** stop condition met **do**
4:    perform variations on the *population*
5:    Evaluate the updated individuals in *population* using the fitness function
6:    Select a number of members to remain in the *population*
7: **end while**
8: **return** *population*

EAs are able to produce high quality solutions via different initial populations, and they can be guided towards creating content with specific desirable features by carefully designing a fitness function. As such, they have been used to generate a variety of types of content, which in addition to levels and maps [159, 207] includes game rules [76, 208], missions [85], puzzles [34], and CCG decks [47]. However, there may be situations where generating a diverse set of generated content is as important as generating high-quality content. For example, imagine you're generating decks for a collectible card game. You would likely want to generate high-performing deck (i.e., able to win games), but you may also want to create different decks to support different play styles [45]. In these cases, we can leverage a Quality-Diversity approach.

2.4.2   Quality-Diversity PCG

Quality-diversity (QD) algorithms differ from standard search-based approaches in that QD approaches try to find a set of solutions with the most coverage over some defined metric space. While exploring the search space, QD algorithms retain a population of solutions, and ensure both the diversity of that population and the quality of the individuals. In addition to a standard and evolutionary algorithm's search space and fitness function, QD algorithms also require encoding of designer knowledge in the form of a **diversity component**. QD algorithms are defined by their:

- **Diversity component**: a method for comparing the similarity of two individuals in a population, and driving the search procedure towards diverse solutions
- **Quality component**: a method for evaluating the fitness of an individual, and pushing the search procedure towards high-quality solutions. This is equivalent to the fitness functions of evolutionary algorithms.

Note the diversity component measures similarity using different metrics than the quality component uses to measure fitness of the individuals, (e.g., evaluating fitness of a maze based on if it is solvable, and evaluating the diversity based on the length of the solution path or the size of the maze).

For a concrete example, we return to the weapon generation problem from earlier. First, let's add another feature to the representation, a weapon type (e.g., sword, hammer, spear, longbow, etc.). Let's say that each of these weapon types have different limits on their damage and time to attack (e.g., a hammer has a higher damage capacity than an sword, but also a higher time to attack range). We will define our quality component to be the initial fitness function proposed,

$$quality(weapon) = 2 \times weapon_{damage} + weapon_{time\ to\ attack}.$$

Now if we only use this fitness functions during evolution, we might find that it is easiest to optimize using a specific weapon type. In our case, however, we want to get high-quality weapons of a variety of types. And so we also introduce a diversity component. We will simply use:

$$diversity(weaponA, weaponB) = \begin{cases} 0 & \text{if } weaponA_{type} = weaponB_{type} \\ 1 & \text{otherwise,} \end{cases}$$

which says that two members are considered different if they are of different weapon types. Now, if we use a QD approach like MAP-Elites [129], then it would start by initializing bins for each of the different weapon types. It would then follow a standard evolution pipeline, of initializing a random population and then evaluating the fitness of the members of the population. Here is where we diverge. Next, MAP-Elites would determine which of the bins each of the members of the population should belong to. If that bin is empty, then the member is stored there. If the bin has a member (or some number of members) in it already, then the new member is only added if it has a higher fitness than one of the existing members in that bin. Then the population is expanded in the standard ways (e.g., mutation, crossover, etc.), and the cycle repeats. At the end of the algorithm, we are left with separate bins for each weapon type, each filled with the highest fitness candidates for that type.

It's important to note that MAP-Elites is just one QD approach, and there are several others that rely on varying search approaches, fitness measures, and diversity components. Gravina et al. [53] survey the topic of using QD approaches for PCG in more detail if the reader is interested.

2.5 Takeaways

In this chapter we offered some brief introductions to the various classical procedural content generation approaches. We discussed how designer and domain knowledge need to be encoded differently for each family of approaches, and discussed some examples of how these approaches could work for different types of content. As you were reading through this chapter you may have thought *"Hmm, I wonder if we could learn some of that encoded knowledge from examples."* Well, in some cases we are able to, at least to a certain extent. And this introduces a core trade-off between classical PCG and PCGML approaches. Classical approaches require the designer to explicitly encode their knowledge to inform the algorithm (either through defining the rules, search space, quality measures, etc.). This gives the designer a huge amount of control over how the PCG approach functions and the created content, but it requires a lot of domain knowledge and expertise from the designer. Alternatively, PCGML approaches (as we will see in the remainder of the book), enable the designer to extract some of the requisite domain knowledge from example pieces of content. This empowers designers that do not have the requisite domain knowledge or don't know how to encode their domain knowledge for a classical PCG approach to still leverage PCG. However, this extraction often does not allow for the same level of direct control as classical PCG approaches (e.g., control over the representation, rules for generation, patterns learned/represented, etc.). Thus, when choosing a PCG approach (either ML-based or classical) the designer needs to consider the tradeoffs and try to select the approach that most closely aligns with their needs.

In Chaps. 3 and 4, we will give an introduction to machine learning basics and theory as they apply to PCG, and discuss some of the ethical and practical considerations around PCGML. Then, in Chaps. 5–10, we will walk through the various families of machine learning-based PCG approaches, and discuss their limitations and applications. Many of those ML approaches will even call back to or support certain classical approaches presented in this chapter.

An Introduction of ML Through PCG

3

In the previous chapter, we discussed various forms of non-Machine Learning based Procedural Content Generation, but before we dive into Procedural Content Generation via Machine Learning (PCGML), we must first discuss what Machine Learning (ML) even is. At its core, ML is a set of techniques for learning functions—a process of finding a mapping from input to output. In the case of PCGML we are trying to learn a function that will procedurally generate content (i.e., map from some input space to some generative space). In this chapter, we will be discussing the basics of machine learning—defining a problem domain, determining the form of function we want to learn, learning the parameters of that function, and assessing the results. In the following chapter we will discuss practical considerations, but for now let's walk through a simple example.

Let's say we have a simple tile-based 2D platformer and we want to make a PCGML level generator for it. We have a set of example levels that we like, and we want our generator to generate similar levels. For instance, let's say one of our example levels looks like Fig. 3.1. This is obviously a very simple level—all we have is empty space and ground of varying heights. There are no enemies, no powerups, we don't even have any platforms! But let's start with this—we need a function that will be able to generate similar levels. Here, we come to the first question of any PCGML approach:

> How to frame a problem so that we can learn from input → output that
> results in a useful generator?

The naïve framing would assume that our output is the level, but there is no clear input that lets us learn this function easily. Instead, we will need some way of constructing the level a little at a time, with the result being the final level. Later, in Sect. 9.3 we will come to approaches where the output is the level, but these are much more complicated than we want to delve into right now).

© The Author(s), under exclusive license to Springer Nature Switzerland AG 2022
M. Guzdial et al.. *Procedural Content Generation via Machine Learning*.
Synthesis Lectures on Games and Computational Intelligence,
https://doi.org/10.1007/978-3-031-16719-5_3

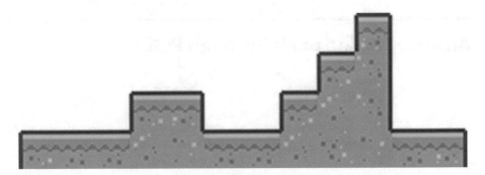

Fig. 3.1 A small sample platformer level. Assets taken from the Kenney Pixel Platformer Pack

3.1   Data and Hypothesis Space

If we take a look back at Fig. 3.1, we can see that the only thing that changes across the level is the height of the ground columns. If this is difficult to see, Fig. 3.2 should make it even clearer. Therefore, we can use ML to predict the height of the next ground column to be able to generate more level one column at a time. We will return to this concept of generating a little bit at a time, as it can help ensure a coherent output. E.g., we want to learn a mapping of INPUT $\rightarrow$ y-value of ground. However, we still don't know what we want to use as our input. For a first pass, let's say that we will use the x-value to predict the y-value. Note, for individual data points we will use the lowercase x, y and for the sets of data we will use the uppercase X, Y. We now need to turn our level into the requisite pairs of input and output data (also known as the **labels**).

When we have a known mapping of input to output, this is known as **Supervised Learning**. However, in **Unsupervised Learning** we do not have this information—which we will discuss later in Chap. 6. Unsupervised Learning is generally focused on finding patterns within the data, without explicit labeling of output. This can come in the form of clustering—where data is grouped together based on the **features** (a characteristic or property of the phenomena that we are capturing in data—for instance, features of a character might be health, mana, stamina, level, etc.) to find similarities and contrasts—or it can come in the form of finding some latent representation of the data that captures features implicit within the explicitly known features. However, for both Supervised and Unsupervised approaches, we need data to learn from. To get data for this toy example, we will look at the height values for the ground in each column (see Fig. 3.2 for a visual representation). This data would look like Table 3.1.

In this toy example, we will treat it as a Supervised learning problem—we have a horizontal position x and we want to predict the corresponding height value y. If we were to graph these points, it would look like Fig. 3.3. We know that these map back to the ground

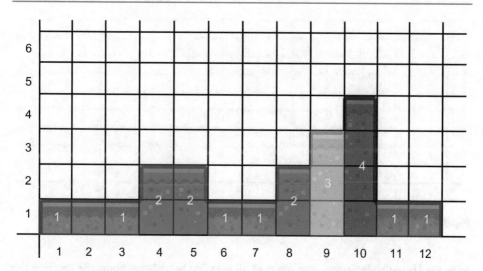

Fig. 3.2 The height values for the ground in the sample level. The maximum height is recorded for each column, since all platforms rise from the ground and there are no floating platforms

Table 3.1 The conversion of our level into data that we can use to train a model	x	y
	1	1
	2	1
	3	1
	4	2
	5	2
	6	1
	7	1
	8	2
	9	3
	10	4
	11	1
	12	1

as seen in the previous figures, but moving forward we will be considering them solely as pairs of input/output.

Now, to learn a function that can predict these values. Let's start with one of the simplest functions—a line. A line is represented mathematically as $w * x + b = y$, where w is the slope, and b is the intercept. We are trying to learn a function that predicts these values (in this case heights), $f(x) = w * x + b$. There are infinitely many possible lines that could be learned by setting the values of the parameters w and b, and the set of all of these is referred

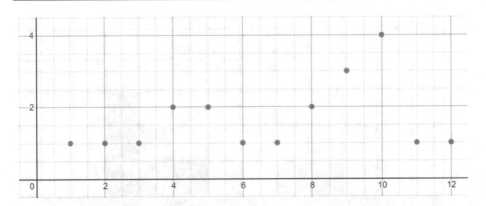

Fig. 3.3 The heights of the level seen in Fig. 3.1

to as the **Hypothesis Space**—the space of all possible hypotheses about the function that maps from input to output. Of course, choosing a linear relationship is just one possible choice of Hypothesis Space—there are infinitely many possible hypothesis spaces that we could choose from (e.g., quadratic polynomial, degree n polynomial, sinusoidal, exponential, logarithmic, etc.)—one of our key tasks when applying PCGML will be to determine which of these hypothesis spaces make sense given our data and the relationship between input and output.

Given the choice of hypothesis space—in this case, a line that predicts a height value for a given horizontal position—we need to find the best member of that hypothesis space. That is, we want the function that does the best job of predicting our height values given the input data.

Figure 3.4 shows a number of different possible lines from our hypothesis space. Our intuition is that some of these lines are better than the others—the solid line always predicts too high, while the dotted line seems to slope in the wrong direction—but how do we quantify this in a way that allows us to learn the best model? The key is that we need some way of quantifying how wrong a model is, this is referred to as choosing a **loss function** (also **loss criterion**, **criterion**, or **objective**): $L(f, X, Y)$, which scores our function f on how well it can use the input X to predict the labels Y. The result of the loss function is commonly just referred to as the **loss** of the function. As a note, not every machine learning algorithm explicitly has a loss criterion; however, every machine learning algorithm either implicitly or explicitly works to minimize or maximize some criterion. Sometimes these criterions are underspecified—many different criterions would lead to the exact same behavior—and sometimes they explicitly specify the exact behavior of the machine learning algorithm. For instance, while loss criteria will not be present in Chap. 6, the probabilistic models have the implicit criterion of trying to maximize the likelihood of the model given the data.

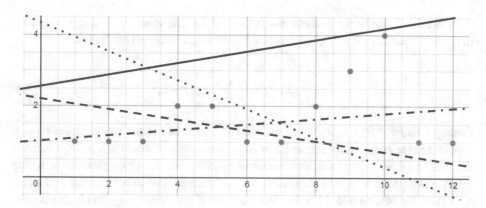

Fig. 3.4 Different lines from the hypothesis space of $f(x) = w * x + b$

3.2   Loss Criterion

We can imagine many different possible loss functions—but at the basic level, we want it to tell us how wrong our model is. The simplest possible loss function that we can imagine is the **0-1 loss**—we look at each prediction and add 1 if it is wrong, and 0 if it is correct. In general, we will sum the loss over our entire dataset:

$$L_{0-1}(f, X, Y) = \sum_{x \in X, y \in Y} 1 \text{ if } f(x) \neq y_i \text{ else } 0 \qquad (3.1)$$

The goal of most machine learning algorithms is to minimize the loss, with the ultimate goal being a model that has a loss of 0 (meaning it perfectly predicts the data). In practice, this is often impossible given the specifics of the data, the hypothesis space, and the loss function. In this specific case, this loss function has some issues—first and foremost being that it does not tell us anything about how wrong we are, just that we are wrong. Look at the lines in Fig. 3.4—none of the lines pass through any of the dots, so all of them would have the exact same loss value—12, because they get all 12 points wrong. However, we have an intuition that some of the lines are better than the others, so let's consider a loss function that helps capture that intuition. For instance, maybe we believe that the distance between the prediction and the true value is something we want to minimize. Just as with the 0-1 loss, this will be 0 if all of the predictions are correct. However, now the model will be penalized more the further away it is from the truth. This will penalize some of the above lines more than the others, giving us more information about which model is the best. We most commonly use the **Mean Squared Error** (MSE) as our loss value for **real** (any value between $-\infty$ and ∞) valued predictions like this—this is also known as L2 or Euclidean loss. MSE computes the difference between the prediction—$f(x)$—and the true value—y, and then squares it (making it a positive number), and finally takes the mean across all datapoints.

$$L_{L2}(f, X, Y) = \frac{1}{n} \sum_{x \in X} (f(x) - y)^2 \tag{3.2}$$

Finding a line that tries to minimize the MSE is known as **Least Squares Regression**, a type of **Linear Regression**. Linear Regressions are a class of machine learning functions that predict a linear relationship between the input and output. In our simple example we are learning two parameters—the parameters being a slope applied to our input w, and a bias b. Without the bias term, our line has to go through $(0, 0)$, which might not be applicable, so we add a bias term to allow the line to pass through any point.

At this point, you might be asking yourself "How do we actually find the slope and bias that minimize the MSE?" For the linear models, the slope and bias are the only **parameters** that we are learning, but other models might have different parameters. Moving forward, we will refer to these parameters as **weights** (so named because they allow the model to *weigh* different kinds of input to varying degrees). Luckily for us, there is a closed-form solution— one that uses a finite number of mathematical expressions—for Least Squares Regression. We can represent our input X and output Y as matrices where each row corresponds to a data point and the columns of X correspond to our features. Then we are trying to find the weights W that solve:

$$Y = XW \tag{3.3}$$

Which we do by finding the weights W that minimize:

$$(XW - Y)^2 \tag{3.4}$$

Which we can solve by finding the point where the derivative with respect to W is equal to 0:

$$0 = -2X^T Y + 2X^T X W \tag{3.5}$$

Which can finally be solved to find:

$$W = (X^T X)^{-1} X^T Y \tag{3.6}$$

The results of which are $w = 0.09$, $b = 1.08$ and can be seen in Fig. 3.5. This does roughly capture the relationship, but given that this is just a line with two parameters, all we can do is find a line, but we don't want our level to just be a straight line (at least, not most of the time). Our model is **underfit**—it does not capture the relationship between input and output found in our data. Our input only includes a single **feature** at this point—the x-position. In

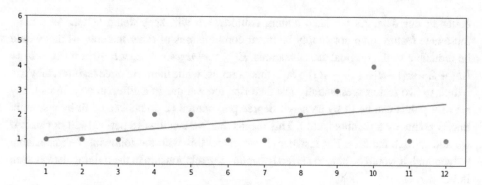

Fig. 3.5 The result of the Least Squares Regression for height values for the simple levels based on horizontal position

this case, we need to determine other features that we could use to help our function learn the underlying phenomenon.

As mentioned earlier, it can often be helpful to generate the level a little at a time. One option would be to look back at what has previously been generated—in this case, use the previous tile's height to help predict the next tile's height. Of course, we need to figure something out for the first tile, since it doesn't have a previous tile. In this case we just set it to 1. The resulting regression solving the equation:

$$y_i = w_x * x_i + w_{y-1} * y_{i-1} + b \tag{3.7}$$

can be seen in Fig. 3.6.

We can tell that the addition of the previous height has improved the performance, although, we do see that there is a lag in the heights. We can keep adding features to

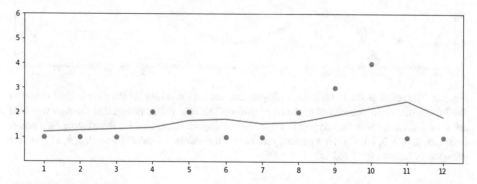

Fig. 3.6 The result of the Least Squares Regression for height values for the simple levels based on horizontal position and previous y value

better fit our data. As we keep adding features, we will keep doing better. As a note, these new features can not simply be linear combinations of other features, as that would be reducible to the original set of features. E.g., $y = w_1 * x + w_2 * b + w_3 * (2x - b) = (w_1 + 2 * w_3) * x + (w_2 - w_3) * b$. In this example, while there are three features, they can reduce to two features, essentially just factoring the weights in a different way. In fact, any set of n points can be fit by an $n - 1$ degree polynomial (2 points can be fit by a straight line, 3 points by a parabola, etc.). This means that we can always perfectly fit our data if we add enough features. For instance, if we model this with the following 7-dimensional polynomial, it would be able to perfectly capture the relationship in the training data as seen in Fig. 3.7.

$$X = [x, x^2, x^3, \ldots x^7] \tag{3.8}$$

Up to this point, we have considered all of our data as **training data**—the data used by the machine learning algorithm. When doing PCGML we often want to use all of the data available to us to train our model—generally speaking, the more training data that we have, the better our end model will be. However, we often want to know how well our model will generalize to unseen data. To do this, we need to reserve some data to **validate** our model—**validation data**. Validation data is most commonly used to assess how much our model is **overfitting**—learning the patterns found in the training data but not generalizing to the unseen validation data.

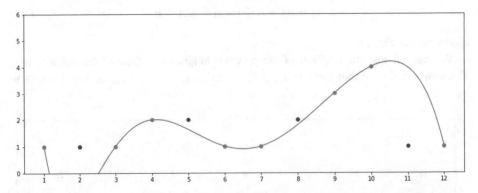

Fig. 3.7 The result of the Least Squares Regression for height values for the simple level based on the 7-dimensional polynomial of horizontal position. The blue points denote that the data was held out for validation, while the red points are in the training set. Note that for the training data points (red: 1, 3, 4, 6, 7, 9, 10, 12) it is a perfectly fit, but for the validation points (blue: 2, 5, 8, 11) the line does poorly (especially 2 and 11)

Although, this comes with a drawback. We have perfectly captured our training data, at the cost of hurting the ability for the model to generalize to unseen data. If we were to evaluate our model for future values 13 and 14 we would get $f(13) \approx -10$, $f(14) \approx -41.4$.

While this model is perfect for the training data, it has **overfit** to that training data. This is one of the key tensions in machine learning—as we add more features and parameters, the model does a better job of predicting the relationship between input and output found in the training data. However, as we add more features and parameters, it is more likely that we are going to overfit and fail to generalize to unseen data. This will become especially true for larger, more complicated models (such as those that are discussed in Chaps. 7, 8, 9), due to the relative lack of training data that will always exist (compare the amount of videogame levels, characters, etc. to the number of sources of text or images that can be found on the internet).

3.3   Underfitting and Overfitting/Variance and Bias

In general, underfitting and overfitting is a problem for all of machine learning. The error of a machine learning model can be decomposed into the error due to **bias**—arising from limitations of the model's capability to learn the relationships (aka **underfitting**)—and **variance**—arising from small changes in input leading to large fluctuations in the output (aka **overfitting**).

The **bias** is defined as the difference between the **expected value** (the mean value over all of the data denoted as $E[\ ]$ with the square braces used to denote that it operates over the distribution of all values) of our model $f(x)$ on unseen data and the true value y:

$$Bias(f, x, y) = E[f(x)] - y \tag{3.9}$$

The **variance** is defined as the expected value of the square difference between the expected value of the model on unseen data and the value of the model on unseen data. In other words, how much the predictions of the model differ from the mean of the predictions:

$$Var(f, x) = E[(E[f(x)] - f(x))^2] \tag{3.10}$$

Looking at Mean Squared Error loss, the loss of our model is:

$$E[(f(x) - y)^2] \tag{3.11}$$

which can be reduced to:

$$(E[f(x)] - y)^2 + E[E[f(x)] - f(x))^2] \tag{3.12}$$

or in terms of Bias and Variance:

Table 3.2 Training and Validation errors for three models

Model	Training MSE	Validation MSE
Linear	0.97	0.54
Quadratic	0.93	**0.38**
7-dimensional	~ 0	3.27

$$Bias(f, x, y)^2 + Var(f, x) \tag{3.13}$$

This means that our error can be broken down in terms of Bias and Variance. Assuming our loss is fixed, there will always be a tradeoff between Bias and Variance. We want to minimize the total loss, but in practice, as we decrease the Bias of the model, we end up increasing the Variance of the model (as seen with the 7 dimensional model above). Conversely, we can reduce the variance, but this often leads to higher bias (as in the first linear example).

Combatting underfitting is easy, we simply add more features to our data (perhaps synthetically), but combatting overfitting is more difficult. However, there are a number of approaches that we can take to assist in this process. The first is holding out some data that we will use to validate our model (as seen in Fig. 3.7). Given that we want our model to generalize to unseen data, by holding some data out of the training process, we can then assess our model's performance on this unseen data. This held out data is referred to as a **Validation Set** (in contrast to the **Training Set**)—and is commonly somewhere between 10 and 30% of the original dataset. We can then train multiple different models on the training set and determine which one is the best by how well it predicts the validation set.

For instance, trying a linear model, a quadratic model, and the 7 dimensional model after holding out points 2, 5, 8 and 11 leads to the mean square errors in Table 3.2. In this case, we would use the quadratic model as it has the lowest error on the unseen data—orders of magnitude less than the 7-dimensional model, despite the fact that the training error for the 7-dimensional is 0 and for the quadratic is 0.93. Of course, intuitively we know that we can do better. The quadratic model still severely underfits the data in this case, so in the coming chapters we will introduce models that are more capable while still not overfitting the training data.

Beyond reserving unseen data for validation, there are a number of **regularization** techniques (discussed in later chapters) that limit the capabilities of the model, hurting their performance while training, with the hope that it will lead to better generalization on the validation data (and beyond). These techniques range from introducing penalties for using more parameters, penalties for larger parameters, randomly turning on and off parts of the model during training, and more.

Returning to procedural content generation, we generally are much more concerned about overfitting than underfitting—we don't want a model to memorize the training data, as it won't generate anything new. However, underfitting is an issue as well, as the content generated by an underfit model will be of lower quality—levels might not be playable, text might be incomprehensible, images might be garbled, etc.

3.4 Takeaways

In this chapter we walked through a toy example of level generation to discuss many of the basic principles of machine learning. Specifically, we discussed the choice of hypothesis space (in our toy example, equations to predict the height of a column); the choice of loss criterion (in our toy example, the mean square error between the true and predicted heights); and a brief discussion on under and over fitting models to data. As previously mentioned, this is a form of **Supervised** learning, we expressly had an input and output. However, this form of turning an unsupervised learning task (learn how to generate new levels that look like the input levels) into a supervised task (learn how to predict the height of the next tile) is sometimes referred to as **Semi-Supervised Learning**. In later chapters we will discuss other types of functions that we can use for this task. This chapter mainly dealt in the theoretical underpinnings of Machine Learning (in the context of PCG), but the next chapter will dive into an overview of the general process of PCGML from data acquisition, through training, to evaluation of content.

So let's say you want to use PCGML to generate something, which seems a reasonable assumption if you're reading this book. In this chapter we'll overview a high level process for doing just that, discussing practical and ethical considerations as we go. We break this process into four steps: (1) produce or acquire training data, (2) train the model, (3) generate content from the model, and (4) evaluate the output content. We'll first overview these steps as a form of introduction then dive into each of them in detail for the remainder of the chapter. We present a visualization of this process as a reference in Fig. 4.1.

In the typical case, you're going to need something to train on, thus our first step is to **Produce or Acquire Training Data**. In most cases this means that we'll need some existing data. There are a number of choices we need to make when acquiring training data, including exactly what content to use, how to represent that content, and so on. For example, let's say we wanted to generate video game music. We'd have to decide what kind of music we wished to generate, do we want something more chiptune or more orchestral? We'd also need to decide what specific songs to use as training data, whether to use whole songs or just parts of songs, how to represent them, and so on.

There are approaches where instead we'll need to define what quality content looks like (such as with PCG via Reinforcement Learning, which we'll cover fully in Chap. 10). Since we're designing the measures of quality, we need to make decisions around what we value and what we don't value for the type of content we plan to generate. For example, we might want to enforce a lack of plot holes in a generated narrative, but may not care to specify anything about the total number of scenes. Or maybe we want to ensure we have at least some minimum number of scenes and less than some maximum, in order to control the output narrative length. These choices are often complicated and non-obvious at the beginning of a project.

Once we have our data, we next need to **Train the Model**. While we say "next," in reality the choices around data and model/training process often occur in tandem. Still, for the

M. Guzdial et al., *Procedural Content Generation via Machine Learning*,
Synthesis Lectures on Games and Computational Intelligence,
https://doi.org/10.1007/978-3-031-16719-5_4

Fig. 4.1 A visualization of the process we identify for this chapter for developing a PCGML generator

purposes of this chapter, we'll discuss these in sequence. For choice of model we often have to design a particular architecture, for example a Markov Model (which we'll discuss in Chap. 6) or a Deep Neural Network (DNN) (Chap. 7). From there, we need to decide how to employ the data to train the model. For example, we can train a DNN via backpropagation (as is typical) or through neuroevolution [44], which is a more niche approach that can be beneficial in certain circumstances. These choices directly impact what can be learned from the available data, and what possible content we can expect as output.

Once we've trained a model on our data or quality measures, we next need to actually **Generate Content** with it. This may seem straightforward: just run the model. In some cases, we do in fact just run the model to get new content. However, in the majority of cases, we still have a number of choices to make about how *exactly* to generate content from our model. For example, how controllable do we want the generation process to be? What guarantees, if any, do we want to include (and what guarantees *can* we include) about the generated content?

The final step of the process is to **Evaluate the Output**. In an academic framework, evaluation is an important step as it allows researchers to determine if their PCGML model supports whatever theory or hypotheses they're exploring. By hypothesis here we mean the classic scientific sense of the term: a proposed explanation for some observed or predicted phenomena. But we're using a broader sense of the word evaluation than typical scientific investigation, including player or user experience. From this standpoint, we can state the purpose of this step as determining whether your PCGML model has achieved your goals, even if your goal is just to generate good content. "Good content" is of course a subjective matter, and difficult to define clearly. Therefore, we often instead use some other measure to approximate goodness or quality, such as user/player reactions or ratings.

Now that we've introduced these four steps, the remainder of this chapter will dive into each of them in more detail, identifying both practical and ethical considerations one should consider during the development process.

4.1   Produce or Acquire Training Data

For most applications of PCGML you'll need data to train your model. The choice of data is important, as it will directly impact what kind of model we can train, and therefore, what kind of final content we'll be able to generate. Because of this, we identify this step as the first and primary one in our process, as visualized in Fig. 4.1. However, one might consider there to be an implicit step 0 to this process in terms of deciding what you want to generate in the first place.

There are many different approaches for producing or acquiring training data, but we can basically split the strategies into two groups. First, you can make use of existing training data.[1] For example, levels, sprites, quests, or sound effects from existing, previously published games. Second, you could instead create the training data yourself. This requires more effort than the first option in most cases, but also allows for more control. This second option can look like explicitly authoring levels, sprites, quests, sound effects, and so on. As a reminder, the core idea of PCGML is to attempt to learn the design knowledge one would otherwise need to hand-author in classical PCG. The benefit then is to sidestep or shorten the process of authoring and tweaking a generator until it can output desired content.

Rather than either of these approaches, one could author an environment that defines what good and bad content look like, where the training data is then produced when the ML model interacts with the environment, as in the case of methods like reinforcement learning. But this is unusual in PCGML for now, and so we leave this discussion to Chap. 10. Below, we discuss some of the considerations that might come up when applying the two different strategies for supervised PCGML.

4.1.1   Existing Training Data

One of the benefits of PCGML is that we don't need to do the same kind of work as when hand-authoring design knowledge for classical PCG or creating all the content ourselves. One might then assume we should always make use of existing training data in order to further reduce our workload. However, there are a number of potential issues to consider when taking this approach.

Author Permission

Existing game content may not exist in an easily-parseable format. While things like the soundtrack of a game may exist as separate files for each song, in most cases the sprites, items, quests, or maps/levels of a game will not. If they are available, then it's likely due to the work of fan communities, for example those who extract or recreate game sprites for classic games. While this work is laudable as cultural expression, and as a way to preserve games, it is not always legal or ethical. We note that we draw a distinction between what is legal and what is ethical, though there is overlap between the two.

Reusing existing game content is at times legally murky, and companies regularly take legal action against individuals who do so. For example, Nintendo regularly copyright strikes fan games making use of game art or mechanics from Nintendo games like Metroid and Pokémon. While the issue of whether using something as training data violates copyright is not yet settled, there is a legal risk in using content from existing games as training data.

---

[1] We include references and links to existing, publicly available datasets in Chap. 13.

In terms of ethics, the fans who originally scraped or recreated the content likely did not do so expecting their work to become the input to an ML model. While this may legally fall under fair use, depending on your PCGML use case (i.e., is this research or an otherwise nonprofit project), individuals may feel that their work is being co-opted or that they are being involved in something against their will. Machine learning itself has become associated with big tech companies who often undertake ethically dubious activities, and so for many individuals any use of ML is tainted by association. For these reasons, when possible, we recommend always getting permission from all individuals involved in authoring/scraping the existing content before using it as training data.

Replicating Existing Bias

The primary limitation when training a model on existing data is that you will end up with output like the existing data. This is, of course, the reason to train on existing data in the first place. If we want a new monster that looks a bit like a Pokémon, it makes sense to train on existing Pokémon. If we want a song that sounds like something from a Sonic game, we could train on Sonic game music, and so on. However, this means that in most cases, training on existing data will not help you achieve a unique aesthetic, you'll instead end up with something that looks/sounds/plays like your existing content. Even this is just the best case scenario, as sometimes your model won't be able to learn to produce anything but noise or will only learn to exactly replicate the training data. We'll discuss some currently open research problems in Chap. 12 that attempt to produce novel content by training on existing content from multiple games, but these methods are still experimental. Thus, in most cases at present, training on existing content means replicating the style, patterns, and features of that content.

This limitation of replicating existing style might seem surprising to individuals familiar with the public output of large ML models for generating text and images, which can replicate a wide range of styles. Those large models are able to achieve style replication specifically because they are trained on massive datasets, which is not possible for PCGML approaches using existing video game data. In addition, these models tend to struggle with less common aesthetics, which occur less frequently in their training data, like those one might find in particular video games.

In the cases where you do want to make use of existing data, there are concerns with replicating existing knowledge that may not be obvious ahead of time. For one, there is the possibility of replicating any harmful or abusive patterns present in the original content. While this can be avoided given a thorough review of the content, these patterns can slip by if one is unaware. For example, if we consider the example of using existing Pokémon as training data, early sprites for the Pokémon "Jynx" have been identified as an example of blackface, a racist caricature associated with minstrel plays of the American south. If someone were unaware of this, and used the original Jynx sprites as part of a dataset, they could end up outputting a Pokémon that seemed to evoke blackface. One could also consider the example of the castle levels from the original Super Mario Bros., where each level ends with a reference to a captured princess. This is an example of the damsel in distress trope, which has been identified as an overused cliche and an example of sexism in video games.

To minimize the risk of replicating harmful patterns, it's important to consider your dataset from a diverse set of perspectives. If you are unfamiliar with the concept of blackface or how overused the damsel in distress trope is in games it wouldn't be possible to recognize these issues in your data. As such, it's important to educate yourself on potential harmful patterns in the kind of content you wish to generate.

Less insidious, but still unwanted biases might appear in your dataset as well. Most machine learning approaches will learn to replicate the structure present in the original training data. When that structure is limiting, it can lead to final models that produce output content that is similarly limiting. Consider for example the hypothetical dataset of tiny level chunks in Fig. 4.2. We use level chunks in this example as they represent a small, but potentially useful type of content to generate. However, these same issues arise in all other types of game content.

If we used this as a dataset there are a few structural patterns that our ML model would likely end up replicating in its output, some of which may not be obvious at first glance. For example, the ground tiles with the grassy top only occur along the bottom of each level chunk and only ever in a single row. The blue platform tiles, comparatively, always occur floating in the chunks, and only ever extending horizontally. Therefore, we'd be highly unlikely to naively generate a level that included a 2×2 block of these blue tiles on the ground. Other biases are less definitive, but would likely still impact the output. Just glancing over the examples you can probably see some patterns that are consistent between rows and columns. The snail and frog characters occur 12 times in the dataset, but stand on the ground tiles in 11 of these 12. Therefore, we'd expect a roughly 11 in 12 chance of these entities showing up on the ground in the generated content, and for it to be unlikely or impossible (depending on the model/generation strategy) to see a floating entity or two entities on top of one another. In some cases this might be exactly what we want, but in others it may make certain outputs we'd like to see unlikely or impossible.

Datasets of other types of games media can have the same issues as identified above. For example, a dataset of music with songs that all have the same instrumentation, the same chord progression, or the same beats per minute (BPM). Alternatively, a dataset of game sprites might all use the same skin colour, the same body shape, or the same types of clothing. It's always worth considering what patterns exist in a dataset, in order to better understand what an ML model will likely learn.

4.1.2 Producing Training Data

Given all of these issues around using existing data, you might consider instead producing your own training data, either explicitly through authoring new content or implicitly by defining an environment and a function to define quality content. However, this has other

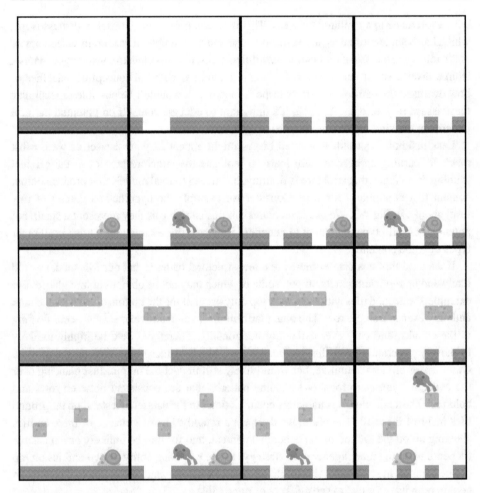

Fig. 4.2 An example, hypothetical dataset composed of 16 level chunks. Assets taken from the Kenney Platformer Pack Redux

issues to consider. For one, if we choose to author new content, there's a question of why we should use PCGML at all. While this varies somewhat between approaches, most ML models require a fair amount of training data. If we find ourselves authoring dozens or hundreds of examples just to train a PCGML model, in many cases we might as well just directly author the content for the game. There are exceptions to this, as we might want players to have an experience that cannot be supported by static content such as a roguelike or one that takes place in a world unique to that player as in Minecraft. If you want to be able to generate millions of pieces of content with your specific style, and it only requires authoring hundreds, that might be worth it! There are many concerns that come up when authoring content, such as the speed of authoring and the quality of the output, but they largely depend upon an individual artist's process and medium of choice, and so we end the discussion of this approach here.

Authoring a function to define what good content looks like may seem like it would take less work, but this is not always the case. While it is hard to author good content, it might be even harder to define what "good" means. For example, maybe we think a good level is one with a lot of challenge, so we measure "challenge" by the number of enemies in the level. But then a generator would just output a level as full of enemies as possible, potentially making it unbeatable. We might change our function to say that levels have to be beatable but still maximizing the number of enemies, but beatable by who or what? If we use another AI agent to play our level it's unlikely to play like a human would. In the worst case, we might end up with a level that's only beatable by one specific AI agent and no one else. We could keep going with this example forever, but will stop here. Suffice to say, academics and practitioners have argued for decades over what good game content looks like. Given that there has been nothing approaching agreement in that time, it's not likely you'll be able to quickly define a "goodness" evaluator function for every PCGML problem. However, writing these kinds of evaluation functions is an art to itself, so with enough experience you can come up with clever ones that can allow you to quickly specify your desired content.

4.2 Train the Model

Once we have the training data we have to train our model. Many of the choices and concerns when it comes to training a particular PCGML approach are unique to that PCGML approach, and we'll cover them in the following chapters. However, there are a couple of broad issues that will arise in most instances of training a model. First, how we choose to represent our data in terms of output size and the representation of the atomic components of our content. For example in Fig. 4.2 we had four types of entities, but we can still represent this data in a variety of ways, as we'll discuss below. Second, how and why to separate data into train, validation, and test splits, which we introduced in Chap. 3.

4.2.1 Output Size

Our choice of how to represent our PCGML problem has an outsized impact as to whether our approach will succeed or fail. The reason for this is fairly straightforward, our choice of representation impacts the complexity of the learning problem. Let's imagine we have the dataset from Fig. 4.2 again. Naïvely, one might assume we should generate a whole chunk of level at once. However, since each chunk is made up of 8×8 tiles, this is equivalent to asking our model to make 64 decisions at once ("what should go in each tile?"). This may not seem too bad, but at every position of this grid we have five possible kinds of tiles (empty, snail, frog, blue floating block, and ground tile). This means that for each one of those 64 positions we need to select one of these five options. So put another way, we're asking our model to select one good level from a space of 5^{64} possible level chunks. Of course we hope to train the best model we can, but we should do everything we can to put the odds in our favor.

One simple option, which we'll discuss further in Chap. 6 is to model a subproblem. For example, instead of generating all 64 tiles at once, we could generate one tile at a time. Looking at the examples, we can see that the bottom left tile is always a ground tile. So if we always started with that tile and generated one tile at a time from there, we could significantly simplify the generation problem. Instead of trying to generate the entire 8×8 we could focus on what tile should come after the ground tile, then what tile should come after that, and so on. Now, a single tile at a time may not be sufficient to capture some structure. For example, we may not be able to create long blue block platforms if we only focus on the tile below (generating going up), since we'd need to see the tile on the left to know where to "line up" the next block. Without this knowledge, the structure of our final output chunks may seem random.

This concept is referred to as **context**—how much context do we need to give the model to make a decision? There's a balance between too little context (random behavior) and too much context (and making the problem too difficult). As a general rule, one can estimate how hard the problem is likely to be in terms of the combinatorics (as we did above) and how many decisions we're implicitly asking the model to make. As long as your training data amount is near this value, your model will likely be able to learn a reasonable generator. In our case, we have 16 pieces of training data, so we likely shouldn't try to make 64 decisions at once, but we can probably support more than one at a time. Despite this rule of thumb, the best PCGML practitioners can do is often to guess and check, as the ultimate determiner of your model's quality will be you.

4.2.2 Representation Complexity

A related question to the size of the representation is how to represent the individual components of a piece of content. For example, consider a game story. We could represent a story in terms of the individual words or in terms of the sequence of events that those words describe. Given the complexity of language, it is likely intuitive which of these would be simpler to model.

In the example from Fig. 4.2 we have five types of tiles, five values that can exist at particular slots or locations of our level chunks. However, this led to a pretty massive number of potential output chunks, and most of these would not be particularly good. To make it more likely for our model to output better level chunks, a common tactic is to reduce the total number of tiles. For example, we could merge the two types of enemies present in this dataset (snails and frogs) into a single "enemy" type. This would help our model generalize over the places enemies tend to occur, as we would go from five examples of frogs and seven examples of snails to twelve total examples of enemies. This may even lead to helpful generalization that improves variety, like generated levels with snails standing atop blue blocks. However, if these two types of enemies are distinct semantically, for example if the frog can jump and so should be the only one off the ground, this may not be ideal.

A loss of semantic knowledge may even lead to unintended readings of the output. Imagine a fantasy game PCGML generator that treated "bandit", "wolf", and "demon" enemies as equivalent. In this case, it could end up placing the first two types of enemies in an in-game equivalent of "Hell", suggesting they have been damned or otherwise sentenced to eternal suffering, which may not be the intended message or befitting the narrative or tone. This same issue can come up across all types of game content: quests ("exorcise a wolf possessing this person"), visuals (representing a bandit running away with a chicken in their mouth), or sound effects (a wolf's howl replaced with demonic chanting). It's important to consider what unintended possibilities might arise from any choice to simplify your representations.

Back to our running example from Fig. 4.2, we could imagine an even further reduction of tile types. We could go from two enemy types to one and two "solid" block types (blue floating blocks and ground blocks) to one, bringing us down to only three tile types total (enemy, solid, and empty). This then would bring our total number of possible outputs to 3^{64} from 5^{64}, an over ten orders of magnitude decrease, without reducing the number of unique outputs from a human perspective. However, if we generate in this smaller, 3-tile representation, we do still likely need to translate back to our five tile representation to present the output to users. Thus, we'd need an extra function to translate between these representations.

One might assume we can simply pick at random in these cases, but this could end with ground tiles in the sky and blue floating block tiles on the ground. This is usually fairly easy to solve, for example with a simple rule that translates a "solid" tile in the 3-tile representation into ground tiles if it's on the bottom of the screen and blue tiles otherwise. These kinds of translation rules are fairly common in PCGML [198], and represent one way to combine classical, rules-based PCG and PCGML. But this does require extra work to put together, and becomes more difficult the more translation is required. For this reason, it's typical to try to strike a middle ground in terms of the representation: too simple and one can lose nuance or make more work for yourself, too complex and the model may not be able to learn.

4.2.3   Train, Validation, and Test Splits

We originally introduced the concepts of train and validation sets in Chap. 3, but here we dive into them in more detail. As a reminder. the concepts of different splits for training data arose in traditional machine learning as a way to ensure that a model had not **overfit**. An overfit model is one that models the training data too closely, to the exclusion of other, unseen data. In PCGML, an overfit model would be able to generate the structures seen in training data, but only those structures, which is not ideal. As such, we can employ heldout data to ensure that we have not overfit, by checking that we can still generate the structures seen in this heldout data. This is the approach we took in Chap. 3. However, we run into a new problem. It's possible that, by using one set of heldout data to double check that we're not overfitting during the development process, that we may end up just overfitting to both this heldout data *and* the training data. To avoid this possibility, we make use of two sets of heldout data. The first set of heldout data is used during the development of the model, and is called the **validation** set or sometimes the development set. The second set of heldout data is used only when we're confident we have our final model, and is used as a final evaluation or test of the model, and is thus called the **test** set.

Our choice for what goes into the train, validation, and test sets is another important one. If we have too little data in the validation or test sets we can run the risk of not recognizing when we've trained a poor performing model. However, in most cases we only have a limited amount of data, and so if the sets are too large, our training data will be too small, and we run the risk of not being able to learn anything at all. A common breakdown is 80-10-10 or 80% of our data going to training data, and 10% each going to validation and test. One approach is to actually resplit the data and retrain the model K times, called a **K-fold cross-validation**. For example, we might split our data into five parts ($K = 5$), each representing 20% of our data, and iterate through each of the folds, using the other four parts as training data. This can help ensure that we didn't just luck out with a particular train-test or train-validation-test split. The exact values and approach will depend upon what is appropriate for a specific use case.

Another choice, which relates to the choice of output size discussed above, is when or where to do the split. If we take a look at Fig. 4.2 again we can see that we have 16 level chunks, which are each 8×8 tiles in size. In this instance, we should likely make use of an 80-10-10 split of roughly 12-13 training instances, and 3-4 train and test instances. However, if we're generating one tile at a time (besides the first tile), we may consider that we actually have 1008 instances of data (63 instances for each of the 16 level chunks since we ignore the first tile). At this point, one might assume we could make use of 80% of these as our training instances, giving us 806 training instances and roughly 100 validation and test instances each. However, if we do this and select these instances randomly from across our dataset, we would run the risk of instances from the same level chunk shared across our train, validation, and test sets. Since each level chunk has repeated structures, this may make it difficult to tell if we've overfit, since the model will in effect have trained on very similar examples across all three sets. Thus, it's generally best to split at the "highest" conceptual level, individual sprites, stories, levels, and so on, rather than at whatever level we're modeling the problem.

Train-validation-test splits are typically only relevant when we're making use of existing data. However, even in the case where we're hand-authoring data, it's helpful to keep these principles in mind. Without intending to, it's possible to construct a dataset that leads the model to learn to generate content that appears good, but that is limited in unintentional ways. For example, we could construct example stories for a generator and find it generates stories that we're initially happy with, but later realize that it only generates output with male protagonists. In cases like this, having the equivalent of a validation or test set in terms of an **inspiring set** of example stories can be helpful. An inspiring set is a set of content you'd like to be able to generate, which you do not use during training time or directly in the generator at all. Instead, we can compare the possible output of the generator to the inspiring set, and make changes to the generator as needed to ensure it can output content like the selected examples. This can help minimize the risk of this or similar issues, though the choice of stories is another place for unintended harmful bias to make its way into your process.

4.3 Generate Content

When we've acquired or produced our training data, determined our representation, and trained our model, we next get to actually generate content. As with training, many of the specific generation decisions will be determined by our particular type of model. However, there's one major question that will come up regardless of the specific model, which is how to balance randomness in the generation process. We can imagine two extremes to illustrate this question. On one hand, we could imagine a cold, logical pursuit of the "best" piece of content, which we refer to as a pure **exploitation** strategy. On the other hand, we could imagine a more free-form, noisy generation approach that ignores anything learned by the model, generating at pure random, which we refer to as a pure **exploration** strategy. In most cases we want to balance between these two extremes, but how to do so is most commonly dependent on the specific PCGML problem and can take many iterations to nail down. However, once we know our general strategy, we will still need to consider any post-processing required to get our content into a workable form.

4.3.1 Exploration vs. Exploitation in Generation

On one hand, we can imagine that we should always generate whatever the model considers to be the best option, either the most probable, the most highly constrained, or the output with the highest weight. But this may not be ideal in all cases. For example, consider the case where we took the most common tile at every coordinate position in Fig. 4.2. Since the most common tile along the bottom row is ground and the most common level elsewhere is empty then we'd end up with a very boring level composed of only continuous ground and empty tiles. If we used this method then we'd never place an enemy tile, since they're always less likely than just empty tiles at the locations where they occur. Thus, this pure exploitation strategy may be unnecessarily limiting.

The extreme alternative would be a pure exploration strategy, where the model ignores everything learned from the data (and the data itself) and just pick a random tile at every position. However, this would likely lead to a noisy and nonsensical mess, and is highly unlikely to match any goals you as a designer have for the output content. Thus, the key is to find a balance between exploration and exploitation.

The typical PCGML solution for the exploration vs. exploitation problem in generation is to make use of **sampling**. In sampling, we make use of the outputs of our model as a way to weight the selection of final content. Let's say we're using a probabilistic model as we'll discuss further in Chap. 6. At a particular position we might have an 80% probability to generate an empty tile, a 15% probability to generate a blue floating block, and a 5% probability of generating a frog. We can then simply generate the final tile based on these probabilities, with a similar approach used by weight or constraint-based approaches. However, the exact sampling methodology will depend on the domain and the desired goal for the model. If we want a model that is more varied, we might push towards the exploration side, or if we want to ensure high-quality output (say if we want to ship the model with a game), we may wish to lean towards the exploitation side.

4.3.2   Postprocessing

Once we have a generated piece of content from our model, the final step is postprocessing. We've already discussed one example of postprocessing, in the example where we reduced the dataset from 5 to 3 tiles, and then would need to have a secondary process translate the 3-tile output levels into the 5-tile output representation. More broadly, postprocessing is an extra step or steps to translate generated content into final output. It can be a transformation of a single output, or the selection of some final piece of content from a set of outputs. The former might look like the translation example above, but it can involve other processes as well. For example, a postprocessing approach could ensure that a piece or set of content met some condition, like removing any stories without a happy ending or a piece of music without a chorus-bridge structure. This process can be done automatically through an artificial intelligence or secondary ML approach, or through human interaction. The second of these can lead to specific problems of its own.

The most common postprocessing issue is **cherry picking**. Cherry picking refers to the process of generating a great deal of content and then only publicly displaying the best results. In some cases, this can be beneficial. For example, for a game we might want to have a limited number of levels, songs, or characters, and so we only pick the best ones to include. However, in the worst case this can give viewers of the cherry picked content an overblown sense of what a generator can output. Famously this occurred with early trailers of the game No Man's Sky, which presented a much more dramatic and varied set of planets and creatures than were present in the game at release. The No Man's Sky developers have since greatly improved this, but at release this led to outrage from players, who felt they'd been misled.

The same issue can arise in the world of tech start-ups or industry labs who have a financial interest in publicly portraying their ML models in the best possible light. However, this can contribute to the public's lack of understanding of what AI and ML can and cannot do, and mislead stakeholders and decision makers. Thus, it's important to consider when to postprocess output and the potential negative repercussions of certain kinds of postprocessing.

4.4 Evaluate the Output

Following this process, you'll end up with a model capable of generating content. In many instances, you'll have begun with a specific goal in mind, e.g., "to train a model that can generate X" or to "produce Y instances of output of sufficient quality." In these cases, determining whether you've succeeded or not is straightforward. However, if your goal is less well-defined or more subjective you may have more difficulty. For example, what if we want to create a generator that produces "fun" output, or create the best generator for a particular kind of content? In these cases, it's worth thinking about how to test whether you've achieved your goal. In other words, how to best structure an evaluation of your system?

In many cases, evaluation of the output is only used as a proxy for understanding the behavior of the generator itself. During development, this can be helpful in terms of determining what to change or tweak to get some desired output, or to just get a sense of what a generator can possibly output. While there are potentially infinite number of ways to structure the evaluation of a PCGML model, we'll overview a few common strategies in this section.

One of the most common ways to evaluate a PCGML generator is to employ a train-validation-test split, and to determine how closely your output content compares to the content in your test set. In terms of this comparison, some PCGML models will allow you to directly predict the likelihood of outputting an instance of the test data, but this is a bit unusual. A more traditional approach is to compare the test set to an equal number of generated outputs from your model. Often a visual inspection is sufficient for informal applications, but if the goal is to prove something about your PCGML generator, it may be necessary to present more quantitative results. In these cases, it's typical to define a set of metrics and to use these to compare the test and output sets [189]. While this does rely on determining a good set of metrics, there is prior research on deriving metrics for many domains, some of which have even been validated with human subject studies [139, 188]. However, metrics alone can be deceptive or lacking context, and so it's typical to include a baseline of an alternative PCGML approach to give a point of comparison.

The second of these common approaches is called an expressive range analysis, which was originally developed by Gillian Smith [172] for non-ML PCG. This is related to the above approach in terms of the use of metrics. However, the goal is to get a broader sense of

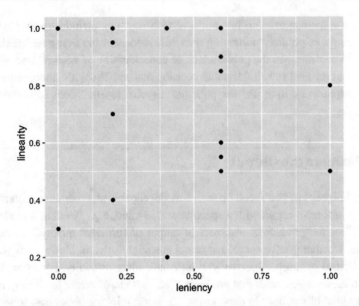

Fig. 4.3 A visualization of the expressive range of the level chunks from Fig. 4.2. Each black dot represents one level from the original set. The x-axis represents the leniency or difficulty of a level, while the y-axis represents the linearity or how flat the level is. This can allow us to get a sense of what kinds of levels exist and (don't exist) in a dataset

the type of output a system can generate, sometimes in comparison to the original input data. For example, we might write an expressive range metric to translate an output level into a value representing challenge or a story into a value representing grammatical correctness.

We can plot levels in terms of these two metrics as in Fig. 4.3, to get a sense of what part of this space our generator covers. In this case, we're using the level chunks used throughout this chapter and plotting the linearity (how well the content fits on a line) and their leniency (how easy or difficult the level is). Linearity in this case was measured as how many tiles existed off a best-fit line, and leniency was determined by counting the number of gaps and enemies. We normalize both metrics to fall between 0 and 1. 0 linearity would mean all tiles fall outside the best-fit line, and 1 linearity would indicate all tiles fall on this line (as in the top row of Fig. 4.2). 0 leniency would mean no enemies or gaps exist and 1 leniency would indicate that five gaps or enemies were present in a level (the maximum number in a single chunk in the original dataset).

We can immediately see that a large portion of the sixteen levels are fairly linear, which is certainly true. But this approach can also help us spot holes in our generated output (pretending in this case that the chunks were generated). For example, no levels have a leniency between 0.6 and 1.0 (in fact there are no instances in the original dataset with a combined total of four enemies or gaps). While this might be obvious with just 16 levels, consider a case where we have dozens, hundreds, or thousands of pieces of content. Being able to quickly identify characteristics of a generator can be a helpful feedback tool for designers, as well as informing them about how to try to tweak the generator to better fit their needs [29]. More recently, Adam Summerville presented an approach for comparing the distributions of the output content over these metrics in terms of the original training or test content, as a means of determining the extent to which the output relates to the original [189].

The final approach is perhaps the most straightforward one to consider, but the most difficult evaluation method to implement: a human subject study. We use the academic term (human subject study), but we could just as easily say a playtest, proofreading, or any other term for testing out some content with an audience. We lack the space to fully cover the practical and ethical intricacies of how to run a human subject study [71]. However it is certainly not as simple as asking someone to play a generated game, read a generated story, or listen to a piece of generated music. In those situations, particularly if an individual knows the person asking, and especially if this is in-person, the respondent may be much less likely to give negative feedback due to social pressure, and a desire not to disappoint.

Getting useful feedback is tricky. Most people can tell you if they liked something or not, but getting more detailed critique can be challenging. For now, we recommend seeking out an expert in human subject studies or referring to a previous example of a human subject study when designing your own. If you have one available to you, an Institutional Review Board (IRB) or ethics review board can be an excellent resource in terms of guidance on how to pursue a human subject study.

4.5 Takeaways

In this chapter we've discussed an abstract process for building a PCGML generator from beginning to end. We hope that this chapter helped give you a mental model in terms of how to develop PCGML generators and some of the issues to consider during that process. Since we have yet to overview any specific PCGML models, we were unable to provide many details, but these will be fleshed out in the remainder of this book. Each of the following chapters will cover some group or class of PCGML models. As you read them, we encourage you to consider how the concepts and issues we identified in this chapter relate to each new approach we introduce.

Constraint-Based PCGML Approaches

5

In Sect. 2.2 we introduced constraint satisfaction approaches to PCG. Recall that constraint-based approaches use a set of **variables**, each with a range of possible values, and a set of **constraints** that must be satisfied over the variables and outputs. In Sect. 2.2 we walked through an example of how we could formulate the problem of generating palindromic words as a constraint satisfaction problem. Constraint-based PCG approaches typically require the encoding of human design and domain knowledge into the set of variables, values, and constraints. For example, when formulating the palindrome problem we encoded the knowledge that the letters at position t and $length - t$ need to be the same and the knowledge of which sequences of letters are allowed into a list of valid words.

In the context of PCG we can imagine encoding the variables to represent sections of levels, and encoding the rules for how different sections could be placed in relation to one another as the constraints. However, manually encoding all the relationships and constraints can be quite a large and error-prone task. What if instead, we had examples of the types of patterns we wanted the constraints to represent? In such a case, we can give these examples of the desired patterns and relationships to a constraint-based PCGML approach. Using a constraint-based PCGML approach we can define the variables and then *learn* the constraints from the provided examples. In the next two sections we will walk through examples of learning design constraints first from platformer levels and then from quests. In the later sections we will discuss a prominent constraint-based PCGML approach, and close with takeaways.

5.1   Learning Platformer Level Constraints

In this section, we show how we can generate a platformer level using a constraint-based approach where the constraints are learned from examples. To start, we need training data. Figure 5.1 (top) shows a section from a platformer level that we will use as our training data

© The Author(s), under exclusive license to Springer Nature Switzerland AG 2022   51
M. Guzdial et al., *Procedural Content Generation via Machine Learning*.
Synthesis Lectures on Games and Computational Intelligence,
https://doi.org/10.1007/978-3-031-16719-5_5

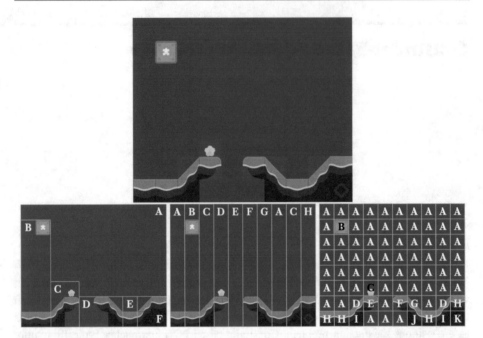

Fig. 5.1 This figure shows a short section of a platforming level (top), and 3 possible representations of that level (bottom). First (left) we have a high-level representation, where regions of the level are the variables and identified structures are used as the possible values. Next (center), we have a lower level representation, where each column is treated as a variable and the unique columns make up the possible values. Lastly (right), we have a low level representation where each sprite position is treated as a variable and the different sprites are the possible values

for this walkthrough. Next, we need to decide how we will represent our problem. Namely, what are the variables and value ranges for those variables? Looking at the example level, we can come up with a number of different representations. A few possible representations are described below, and visualizations can be seen in Fig. 5.1 (bottom):

- a tile-group or structure-based representation where we treat the regions of a level as variables and different structure types (e.g., empty space (A), uphill with item (C)) as the possible values (Fig. 5.1 bottom, left)
- a column-based representation where we treat the columns in the level as variables, and the columns that appear in the level as the range of possible values (Fig. 5.1 bottom, center)
- or a tile-based representation where we treat each position in the level grid as a variable and the possible values are the sprites that appear in the level (Fig. 5.1 bottom, right)

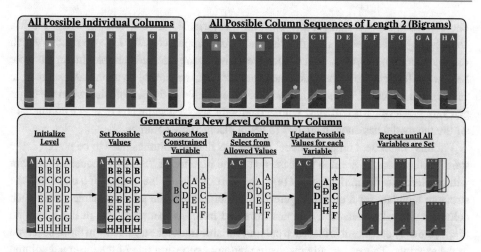

Fig. 5.2 This figure shows an example run of using a constraint-based generation approach using a bigram model, or an n-gram with $n = 2$

In our case, we will use the column-based representation. We chose this representation because it offers a balance of expressiveness and variance. The structure-based representation would be very rigid in the arrangement of the level due to the size of the structures, and the tile-based representation could be very noisy in its output without careful consideration of how to model the relationships. In Sect. 5.3 we show an approach using a tile-based representation, and in Chap. 6 we present several probabilistic PCGML approaches that use a tile-based representation.

Using this column-based representation we can represent a level as a sequence of variables (i.e, col_1, col_2, ..., col_n), where each col_i takes its value from one of the columns observed in the training level. Figure 5.2 (top-left) shows the possible set of values each variable can take. Using this representation, the constraints we want to learn are the possible column sequences allowed in a level. We will use a simplified **n-gram** model here to capture column sequences of length n that are possible given the example level. An n-gram model typically models a distribution over possible sequences of length n, but we will be using it instead to model whether a sequence was observed or not. Chapter 6 will revisit the n-gram model as a probabilistic approach.

Using $n = 1$ (a **unigram**), we extract all the sequences of length 1 that appear in the example level. We can then treat these sequences as constraints, by making it so that only these sequences are allowed in the output. Using a sequence of length 1 won't provide much context to the model, and we will see the consequences of this choice later on. With these constraints learned, we have a fully defined constraint satisfaction problem, and we can use the generation algorithm described in Sect. 2.2. Recall that this constraint satisfaction algorithm starts with a sequence of unassigned variables, col_1, col_2, ..., col_n, chooses one of the unassigned variables, and then assigns the selected variable one of its possible values.

The algorithm repeats that process until all the variables are set, or until a variable has no possible values. Recall also, that there are many approaches for *how* we choose which variable to assign as well as *which* value we assign to the selected variable. In our example here, we will select the most constrained variable (i.e., the variable with the fewest possible values), and select the value randomly from the available options. Note that when we begin generating a level, we will start with a randomly selected sequence of $n - 1$ columns to ensure the model has enough context to begin placing new columns.

For this contextless unigram model this means that no columns need to be placed at the start. Furthermore, it is often the case that some number of "Start of Sequence" content will be placed as the initial context (e.g., the start of a level, the outline of a character, the structure template for a quest). Figure 5.2 (top-left) shows the set of possible values each column could take, using the columns extracted from the example level in Fig. 5.1. With $n = 1$, the only constraint is that each column variable must take a value from those observed in the example. This essentially results in randomly choosing one of the observed column values for each position in the generated level. Randomly choosing columns in this way can lead to some undesirable structures in the output. For example, what if we chose one of these columns 5 or 10 times in a row, or what if we just kept choosing columns A and H alternating?

When $n > 1$, things get a bit more interesting. With $n = 2$ (also known as **bigram**), we extract every pair of columns that occur in sequence in the example level and use them as constraints. That is, the value of each column variable in the level to be generated is constrained by the value of the column to its left. For this example we assume the level wraps around; practically this means that we view the first column A as following final column H, which helps to reduce dead ends during generation. Figure 5.2 (top-right) shows the bigrams extracted from the training level in Fig. 5.1. The rest of Fig. 5.2 (bottom) shows how one example run of the generation process using a bigram model could work when generating a 5 column long level. We start with a randomly selected column value for the starting column variable (which happens to be column A, Fig. 5.2, bottom-first). Next, we propagate the effects of that choice by updating the sets of possible column values for each of the remaining columns variables (Fig. 5.2, bottom-second). Here the second column has its set of possible values reduced to only B or C. This is because in the training level these were the only column values observed immediately following a column of type A. Notice that the remaining columns' possible values are reduced as well by further propagating the constraints according to the possible values in each previous column. For example, the third column can only be C, D, or H. This is because we know column 2 can only be either B or C; and we know B must be followed by C, and C must be followed by either D or H. This process is continued for the remaining columns as well. We then go through the generation loop of choosing the most constrained variable (i.e., the next column to assign a value to), randomly selecting a value for that variable from the possible values remaining, and finally propagating the effects of that choice to the remaining variables (Fig. 5.2, bottom).

Table 5.1 This table show the size of the generative space for the n-gram model with increasing values of n for levels of length 5. We use length 5 here to stay consistent with the generation example in Fig. 5.2, and highlight how different values of n should affect that generation task

n	Unique n-grams	Number of possible levels
1	8	32, 768
2	10	25
3	10	10
4	10	10

When $n \geq 2$, the learned constraints are able to encode ordering and sequencing information from the example level. This introduces the core tension of constraint-based PCGML approaches. As the context available to the model increases (for this model, as n increases), it improves the coherency of the generated content. However, increasing the context (i.e., increasing n) also leads to the model becoming more constrained and lower variety in the output. At the extreme end, where n is the length of the input sequence, then the model can only memorize and copy the exact sequence observed in the input level. On the other side, when the model has less context (i.e., with a smaller n), it is less constrained. This leads to more randomness and variety in the output, but also potentially less coherency if n is not large enough to capture the relevant patterns in the input level. As a reminder, at the extreme where $n = 1$, the model has no context and can only randomly choose between all possible values.

Table 5.1 shows an example of this tension using our training level by computing the number of possible unique levels that can be generated with varying values of n. This was calculated by starting generation with each possible set of $n - 1$ length sequences, and counting how many unique levels those starting sequences of columns can lead to. When $n = 1$, the number of possible levels is simply $UniqueColumns^{LevelLength}$ or 8^5, since every possible column can be placed in every possible position (i.e., there are no restrictions on ordering or using the same column multiple times in a level). When we start providing any context, $n \geq 2$, the model starts to account for sequences observed in the training level and the number of possible levels plummets. Further, notice for $n = 3$ and $n = 4$, the number of possible levels is the same. In fact, at this point the model is already memorizing the training level and replicating a sliding window over that level for the generated level. Notice that every pair of columns observed in the level only occurs once. This means, that the column following that pair (the third column in the n-gram when $n = 3$), will be uniquely associated to that pair. Basically, 2 columns is enough context to uniquely identify what the next column should be; and so, when $n \geq 3$ the model has memorized the input sequences. The reason the number of possible levels is 10 when $n \geq 3$ is because the length of the training level memorized is 10 columns, and each position functions to slide or offset the start of the memorized sequence. This is in part due to our very small training set, which provides little variety to the model. However, as stated above, this memorization will happen to some extent with any training set once enough context is given to the model.

5.2   Learning Quest Constraints

In this section, we walk through another example of using constraint-based PCGML, this time for quest generation. Quests are often smaller missions or side-goals appearing in many role-playing games. In our examples, we assume that quests are represented by a series of tasks that a player needs to perform [98]. An example of one such quest could be,

1. Go to **the cave**.
2. Get the **broken sword**.
3. Talk to **the blacksmith**.

In classical PCG approaches for quest generation, we might use these sequences of tasks as a type of quest template, and treat the text at the **bolded** positions as the variables to be set by the generator. In this case, the set of possible variable values would be the lists of allowed allowed locations, items, and non-player characters (NPCs). With this setup, if we had a list of other locations, items, and NPCs, we could slot them into this template to create a new quest, such as:

1. Go to **the dungeon**.
2. Get the **magic book**.
3. Talk to **the wizard**.

If we're using a classical constraint-based PCG approach the constraints could be defined by a human writing a list of allowed values per variable with no other constraints, or perhaps the designer would define additional constraints on which locations, items, and NPCs could occur in a quest together (e.g., maybe bringing a magic book to a blacksmith should be avoided).

Now, if we are to use a constraint-based PCGML approach for this problem, we will first need some training examples. Figure 5.3 shows four example quests we will use as our training set. We will use the flow of the tasks in these example quests as templates for future quests. For example, the **Hunt** quest template follows the "Go to ⟨location⟩," "Kill ⟨NPC⟩," "Get the ⟨item⟩." We will use ⟨location, item, NPC⟩ as the variables, and extract the observed values of these variables from the example quests to construct the value ranges for these variables. The training stage of this approach is then constructing the possible variable values from the examples. This gives us the following possible variable values:

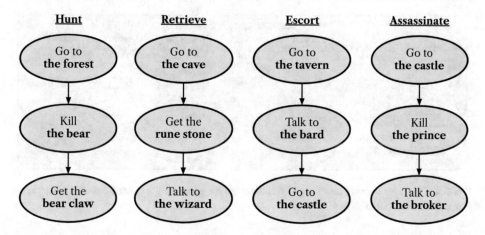

Fig. 5.3 This figure shows a few example quests we can use to train our quest generator

- `item`: [**bear claw, rune stone**]
- `location`: [**the castle, the cave, the forest, the tavern**]
- NPC: [**the bard, the bear, the broker, the prince**]

For simplicity, we will assume there are no limitations on which variable values can be assigned together in a quest.

For generation, we can start with one of the human-authored quest templates from above (i.e., **Hunt, Retrieve, Escort,** or **Assassinate**), and fill in the variable values. Let's choose the **Escort** template, then we need to select the variable values for $location_1$, NPC, and $location_2$. We can follow the same steps as when sampling the platformer level in the previous section: select the variable to set, set the value, and update the other possible values. Figure 5.4 shows one example of generating a new quest using the **Escort** quest template with the learned possible variable values. Note that during generation, since there are no interactions between the variable values, the next variable to set can be chosen in any order.

Generating quests in this way can get us some variation, but the structure always stays the same for each of the templates. Let's extend this approach by modelling the sequence of tasks in the templates in addition to the other variables. For this to work we need to represent the tasks themselves as variables and learn the constraints on the sequences of tasks. From the same training examples above, we can extract allowed sequences of tasks or which tasks can occur immediately after another.

- *Get* can be followed by: [**Talk**]
- *Go* can be followed by: [**Get, Kill, Talk**]
- *Kill* can be followed by: [**Get, Go**]
- *Talk* can be followed by: [**Go**]

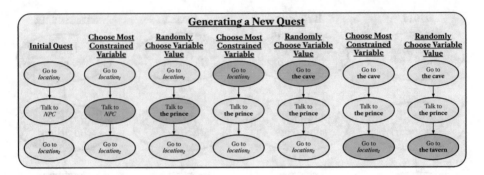

Fig. 5.4 This figure shows an example walkthrough for generating a quest from a template using learned sets of variable values

Figure 5.5 shows which task can flow from another. Notice that this learning step is similar to the bigram modeling for platformer columns; we are modeling sequences of tasks now, but still using only the immediately previous task as the context.

Using these learned constraints, we can now generate new quest templates. We can start by choosing a sequence length for the quest. We then, again, follow the same steps of choosing the variable to set (in this case a task in the sequence), choosing a value from the set of possible tasks, and propagating the effects of that choice. Figure 5.6 shows what this process might look like. We start by choosing a length for the quest, in this case five tasks. Next, we choose which variable to set. We will again be choosing the variable with the fewest options. In the first iteration, all variables have the same number of options, so we randomly select the second task. Next, we randomly choose the value for this variable from the possible values. Here we choose the *Talk* task. We then propagate the effects of setting that value, through the other variables and reduce the set of possible values. We now repeat this process, of selecting the most constrained variable, setting the value, and propagating.

Fig. 5.5 This figure shows a graph representing the constraints for the task sequences learned from the examples

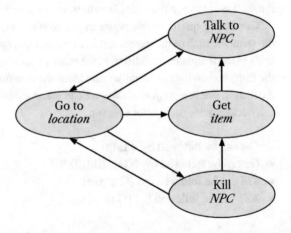

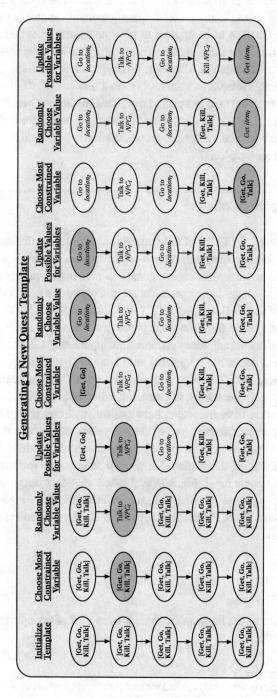

Fig. 5.6 This figure shows how we can generate a new quest template using the learned task sequence constraints. The resulting template can be fed directly into the previous quest generation approach to set the variables within each task

Continuing this process until all variables are set results in a new quest template. This new template can then be passed to the previous quest generation approach to set each of the variables within the template tasks in order to get a new quest from that template. Using this template generation approach, we can get more variety in the templates than when only using the hand-authored templates. However, this approach is not perfect as it can result in strange or undesirable templates, such as, "Go to location," "Talk to NPC," "Go to location," "Talk to NPC," "Go to location." This can happen because the problem is under-constrained (i.e., too little context). Using what you've seen in the platformer example and this example, how much more context do you think you could add to the model before it starts memorizing the templates?

5.3   WaveFunctionCollapse

WaveFunctionCollapse[1] (WFC) is a PCG approach that has been described as both a PCGML approach [195] and as a constraint satisfaction approach [90]. The original implementation of WFC was focused on image and texture generation, but it was quickly adapted to work in other domains (e.g., graphs, language, and other grid types) and incorporated into several games (e.g., *Caves of Qud* [22], *Bad North* [27], and *Townscaper* [187]). WaveFunctionCollapse was quickly adopted in the indie game industry as an easy to use PCGML approach for extracting and replicating adjacency patterns from input examples, which did not require in-depth knowledge of the internal algorithm nor tuning of algorithmic parameters. Games academics took notice of WFC as a new constraint-based PCGML approach that can easily be leveraged without deep knowledge of constraint satisfaction problems and approaches.

WFC essentially uses three high-level functions for modeling and generating content: *Extract*, the training stage; *Observe*, part of the generation stage where it samples a value for a variable; and *Propagate*, another part of the generation stage where the effects of the sampling are spread. During this explanation we will demonstrate each step using a 4×4 pixel image as the example input; Fig. 5.7 (left) shows the input image we will be using, and Fig. 5.8 shows the generation process. In our walkthrough of the algorithm we will be assuming a 2D domain, but WFC has been applied to 3D domains and can be extended to arbitrary dimensions.

5.3.1   Extract

The *Extract* function handles the training stage; it takes a set of example artifacts and pulls out:

---

[1] WaveFunctionCollapse is commonly spelled without spaces when referring to the PCG algorithm to differentiate it from the physics phenomenon of the same name.

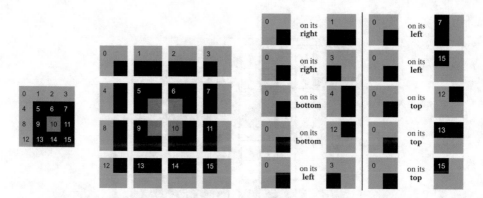

Fig. 5.7 This figure shows how the WaveFunctionCollapse *Extract* function works on an example image. Left shows the input image, and center shows the extracted patterns. Notice that the input image is assumed to wrap around. The right shows the allowed adjacencies for pattern 0 (i.e., which patterns can occur around it without causing a contradiction). Note that we are using the *overlapping* version of WFC, where the extracted patterns overlap one another giving the adjacency constraints (e.g., patterns 0 and 1 share two pixels from the input)

1. the $N \times M$ patterns (e.g., 3×3 groups of pixels, tiles, nodes, etc.)
2. the number of times each of those patterns occurs in the provided examples; it uses these frequencies when selecting patterns as we'll see later
3. the adjacency rules for those patterns (e.g., which patterns can be placed next to one another)

The *Extract* function finds the possible patterns by sliding a window of the desired size over the input artifact, and keeping track of any $N \times M$ window it encounters (or $N \times M \times Z$ for 3D domains). In Fig. 5.7 (center), we can see the patterns found when sliding a 2×2 window over the input image (left). In our case, we are using the *overlapping* version of WFC, and so each of the extracted patterns has pixel values that overlap with other patterns (e.g., patterns 0 and 1 overlap sharing a vertical pair of pixels, and patterns 2 and 6 overlap sharing a horizontal pair of pixels).

We also assume the input image is meant to be wrapped,[2] meaning that when the sliding window extends outside the boundary of the image, it fills the extra cells from the other side of the input image. For example, pattern 3 extends 1 pixel too far to the right, and so we fill that extra column with pixels from the first column of the input image (all grey cells). Similarly, pattern 15 extends too far to the right and too far down, and so we fill the bottom row and right-most column using pixels from the top row and left-most column from the input image. This process results in a total of 16 patterns from the input image; 1 for each

---

[2] Alternatively, we could use the non-wrapped approach. In our case this would mean that that the last column and bottom row of extracted patterns would be removed (i.e., patterns 3, 7, 11, 12, 13, 14, and 15). This would also result in many fewer pattern adjacencies and less variety in the output.

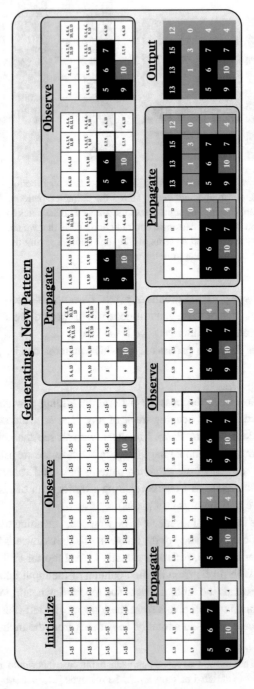

Fig. 5.8 This figure shows the *Observe* and *Propagate* cycle when generating a small pattern with WaveFunctionCollapse

position in the image. However, some of the patterns extracted are in fact the same (1 and 2, 4 and 8, 7 and 11, 13 and 14). In this case we only record 1 pattern type to represent groups of duplicates, and record how many times each pattern type occurred in the input image to inform the pattern selection during generation.

Once the patterns and frequencies have been recorded, the *Extract* function extracts the adjacency rules from the example. Here we are determining directional rules for placing patterns; and this can be learned by directly observing the directional relationships of the patterns in the example. For example, for pattern 5: we know that pattern type 1 or 9 must occur above; pattern 9 must occur below; pattern type 4 or 6 must be to the left; and pattern type 6 must be to the right. These adjacency rules also can be defined by observing which patterns can be placed next to each other without contradictions. In Fig. 5.7 (right), we can see the adjacencies for pattern 0 both directly observed from the input, and those patterns that can be placed without contradicting one another. Pattern 4 was directly observed below pattern 0, and pattern 12 was observed to not contradict pattern 0 if placed below, and so it is also a valid adjacency here. These learned relations ensure that the red pixel will always be enclosed in black pixels in the generated output, which may be desirable or not depending on the domain and designer's intent. For example, an island always being surrounded by shallow water might be fine for the model to learn but a treasure chest always being surrounded by traps might be less desirable. This is just to say that the training data will have a big impact on any learned models, but in particular models which are touted as needing very little training data, such as this one.

5.3.2 Observe

The *Observe* and *Propagate* functions work together to generate new content. We will reference Fig. 5.8 throughout as a example of this process. First, a new $X \times Y$ grid of unobserved cells is initialized, where each cell contains the set of all possible patterns that can be placed. Initially any pattern can be placed in any cell. Figure 5.8 (*Initialize*) shows an initialized grid where each cell can take any of the pixel values (i.e., red (R), grey (G), or black (B)), since any of the patterns can be placed. Then the WFC algorithm alternates between the *Observe* and *Propagate* functions. First, the *Observe* function is used to find the most constrained position in the output grid (i.e., the position with the fewest possible patterns with ties broken randomly). As with previous constraint-based PCG and PCGML approaches, various selection methods can be used here; in fact, Karth and Smith [90] explore the effects of different selection methods for WFC. The *Observe* function then randomly assigns one of the possible patterns available to that cell using the extracted pattern frequencies from the *Extract* function as weights. The first grid in each of the **Observe** blocks in Fig. 5.8 shows the selection of which cell to fill (by highlighting the borders of that cell), and the second grid in the **Observe** shows the result of filling that cell with one of the possible pattern values. Assigning a 2×2 pattern to a single cell in a grid means that the cell takes the pixel value from the top-left position in that pattern; in the case of choosing pattern 10 in the first *Observe* block, this means setting the color to red. At the same time, the rest of the pixels in the chosen pattern reduce the set of patterns for the surrounding cells. Assigning a definite pattern to this cell may have an effect on which patterns are possible at various cells throughout the grid. Therefore, the *Propagate* function is called next.

5.3.3 Propagate

The *Propagate* function can be any constraint propagation function that will use the known and possible values in each cell along with the learned adjacency rules to update which patterns are still possible at each position. The WFC algorithm will then call the *Observe* function again, and so on, until the entire grid is populated. After a *Propagate* call, it is possible that a position will have no possible patterns available, meaning that there is no solution to the current instantiation of the problem. In this case, the WFC algorithm restarts with a new unfilled $X \times Y$ grid.

For a more concrete example, let's return to the first *Observe* block in Fig. 5.8 where the algorithm chose pattern 10 for the given cell, resulting in a red color assignment for the cell. In the following *Propagate* block, the effects of that pattern choice are spread through the grid. Red is an uncommon color in the training patterns and only occurs in specific configurations. As a result, the set of possible patterns to the left and above the current cell are reduced significantly. Namely, they are reduced to the patterns that contain a red tile in specific positions (i.e., patterns 5, 6, and 9). The remaining cells are less constrained, but still have their possibilities reduced quite a bit to accommodate the overlapping sections of the patterns. Notice that the cells further from the observed cell tend to have more possibilities; this is due to the local nature of the patterns and adjacencies learned; in a larger example this would be even more apparent.

WFC essentially uses the AC-3 constraint satisfaction algorithm [227] with the main differences being that WFC learns its rules from data and randomly samples its variable values according to observed frequency. WFC has also been likened to the model synthesis algorithm [123], which also extracts constraints for local patterns and arrangements from examples. The main differences between model synthesis and WFC are that they use different heuristics for selecting which variable to set and the model synthesis algorithm can generate new content in chunks to avoid dead ends. A more detailed discussion on the differences between WFC and model synthesis by the creators of the respective algorithms can be found here[3] and here.[4]

---

[3] This model synthesis algorithm creator's (Paul Merrell) discussion of the differences between WFC and model synthesis: https://paulmerrell.org/wp-content/uploads/2021/07/comparison.pdf.

[4] WFC creator's (Maxim Gumin) discussion of the differences between WFC and model synthesis: https://github.com/mxgmn/WaveFunctionCollapse#used-work.

5.3.4 Extending WaveFunctionCollapse

It is important to note that despite its relatively wide adoption, WFC has its limitations. One core limitation of WFC is that it only learns local patterns, and the scale of the patterns learned are dependent upon the settings of the algorithm. For example, in a platforming level there may be a 4 sprite wide gap. If WFC is only told to capture patterns of size 2 or 3, it is possible that it will produce much wider gaps by placing the middle gap patterns next to each other continually. This will potentially result in unbeatable levels. However, in addition to the base WFC algorithm described above, researchers and practitioners have created numerous extensions.

To increase the controllability of the algorithm, Karth and Smith [89] adapted WFC to accept negative examples or examples of patterns that the designer does not want the algorithm to produce. They discuss how this extension can make the approach more usable as a mixed-initiative tool, and particularly more usable by those without expert ML or WFC knowledge. Being able to explicitly show the algorithm patterns can make the approach easier to use and control. Other extensions to allow for improved control have relied on the fact that WFC uses constraint-solving for generation. For example, Maxim Gumin[5] has shown that WFC generation can be applied to partially filled outputs, which allows users to provide a starting point for the generation and therefore control the output to some extent. Additionally, Karth and Smith [90] have done detailed analyses of WFC from a constraint-satisfaction perspective; and explore using different approaches and settings for each of the stages of the algorithm (i.e., *Extract*, *Observe*, and *Propagate*).

WFC has also been applied to a variety of domains outside of standard grids. A number of people have used WFC in non-standard grid settings, such as hexagonal grids [18], irregular grids [187], and 3D grids [187, 202]. Others have applied the algorithm to completely different domains, such as poetry generation [133], hierarchical level generation [22], and even music generation [10]. These examples show that adjacent positions can be defined by neighbors in a grid, adjacency through sequences, or adjacencies defined through graphs nodes and edges. Essentially, this approach can be adapted to work with arbitrary definitions of adjacency, and the work mentioned above has explored those limits.

5.4 Takeaways

Constraint-based PCG approaches define some generative space by specifying what is allowed in that space using constraints. These rules can be defined manually, as we saw in Sect. 2.2 or learned from examples as we showed in this chapter. Being able to see these explicit definitions of what is allowed in the generative space can open up routes for easier understanding of the models. Constraint-based PCGML is an underexplored area and

---

[5] Using WFC to generate from a partially filled grid: https://github.com/mxgmn/WaveFunctionCollapse/blob/master/images/constrained.gif.

new approaches may appear at any time (much like how WFC appeared and was quickly adopted). Many of the existing constraint-based approaches can be adapted to learning the constraints from examples, by representing the data properly and extracting relevant co-occurrences. Alternatively, many existing PCGML approaches can also be adapted to perform as constraint-based approaches. For example, Opara's approach[6,7] builds on much of the existing texture synthesis work, and in particular the modeling of the neighborhood rules and relationships is handled similarly to Markov Random Fields (MRFs) [31], a model we discuss in the following chapter.

As you read through this chapter you may have wondered whether we can model how common a patterns is in addition to whether the pattern is allowed or not. Well, this is where probabilistic PCGML approaches come into play. If we think of constraint-based PCGML approaches as modeling a "yes" or "no" for patterns in the generative space, then we can think of probabilistic PCGML approaches as instead modeling "how often" for the patterns. Many of the constraint-based approaches learned in this chapter could be converted to probabilistic approaches by capturing the probability of the patterns instead of only the possibility of them. In the next chapter, we will introduce probabilistic PCGML, and we will find that this is a fairly common relationship between constraint-based and probabilistic approaches.

---

[6] Nordic Game Conference presentation: https://youtu.be/fMbK7PYQux4?t=463.

[7] Medium Blog post: https://medium.com/embarkstudios/texture-synthesis-and-remixing-from-a-single-example-faf5f4e8a5b8.

Probabilistic PCGML Approaches

6

In the previous chapter we introduced our first category of PCGML techniques, namely, constraint-based PCGML. Constraint-based PCGML approaches rely on data to build a model representing the structures and relationships that are allowed and not allowed between a set of variables. But what if we want to instead capture how likely or common a relationship is and not just whether it occurs? In these cases, we can start leveraging **probabilistic PCGML** approaches.

Probabilistic PCGML approaches model a domain by capturing how likely different relationships and patterns are from the provided data. In the following sections we give more concrete descriptions of what this means. But conceptually, if we think about flipping a coin, a constraint-based model would say it is possible for the coin to land on either heads or tails. A probabilistic model would tell us that a fair coin will land on heads 50% of the time, and tails the other 50%. In the next section, we revisit generating a platformer level, this time with a probabilistic approach, and use this to introduce different types of probabilities. Afterwards, we discuss some important concepts for probabilistic approaches. We then introduce clustering and its ties to PCGML. Throughout the chapter as we introduce a new probabilistic approach, we discuss prior work that leveraged that approach. By the end of the chapter you should have an overview of what probabilistic PCGML has been used for in the past, and hopefully some ideas for how you could use it!

6.1   What are Probabilities?

Before we jump into using probability to generate content, we first have to define what we mean by **probabilities**. Probabilities represent the likelihood of different events by assigning all possible events a probability value between 0 and 1 where the sum of these values is 1. A **random variable** is used to describe a set of possible events. If the possible events are $\{A, B, C\}$, then the random variable could be $X = \{A, B, C\}$. The probability of the event,

© The Author(s), under exclusive license to Springer Nature Switzerland AG 2022   67
M. Guzdial et al., *Procedural Content Generation via Machine Learning*,
Synthesis Lectures on Games and Computational Intelligence,
https://doi.org/10.1007/978-3-031-16719-5_6

A, is denoted $P(X = A)$ or $P(A)$, for shorthand. Here X is an abstract event, so $P(X = A)$ essentially means: the probability that some abstract event X is the event A. The shorthand $P(A)$ reduces this to just say the probability of event A. Random variables are separate from the probability distributions, but the distributions can help us reason about the random variable and its outcomes. In the simplest example, all possible outcomes have the same likelihood. In this case, the probability of an outcome is equal to the number of ways that outcome can occur divided by the total number of ways all outcomes can occur. This type of distribution is called a **uniform distribution**, because all the probabilities are the same. For example, imagine we have a fair 6-sided die. The random variable describing our die would be $X = \{1, 2, 3, 4, 5, 6\}$, indicating the different sides of the die. Then the probability of rolling each of the values is simply the number of times that value appears on the die, divided by the total number of sides. For our fair die, each side has a $\frac{1}{6}$ chance.

However, in most practical cases, and particularly in PCGML, we are trying to infer the probability of events or outcomes because we believe they can hold useful information. In these cases we can estimate the probability distribution from observations or training data. For example, imagine we have been using a different six-sided die to play a game, but we've become suspicious that it might be weighted or unfair in some way. The random variable would remain as $X = \{1, 2, 3, 4, 5, 6\}$. But now we can construct a probability distribution for the outcomes using observations and see how it aligns with the desired uniform distribution. To do this we would need to roll the die many, many times and record the outcomes. Let's say we rolled the die 100 times and got the results in Table 6.1.

From these observations we could build a probability distribution that describes the die's behavior. And it seems our suspicion that the die is not behaving in a uniform way was correct. For instance, for our fair die we know that $P(X = 5) = \frac{1}{6}$, but from our constructed distribution we see that $P(X = 5) = \frac{1}{100}$! It is important to note here that the more samples we observe, the more confident we can be in the constructed distribution; that is, the more times we observe the different possible events the more information we gain about their likelihoods, and the more confidence we can place in the constructed distribution. In many

Table 6.1 This table shows the results we might observe when rolling a suspicious die 100 times. We have the number of times the die landed on each side (Occurrences), and the associated probability ($P(X = \text{Die side})$)

Die side	Occurrences	$P(X = \text{Die side})$
1	21	$\frac{21}{100}$
2	4	$\frac{4}{100}$
3	22	$\frac{22}{100}$
4	21	$\frac{21}{100}$
5	1	$\frac{1}{100}$
6	31	$\frac{31}{100}$

domains 100 samples is a very small amount (particularly if there are may options, e.g., a 20 or 100-sided die instead of six-sided die). So before you accuse a friend of using weighted die, you should be sure to collect many more observations.

In the next section we will describe how we can use a method similar to the method above to construct a probabilistic PCGML model for level generation.

6.1.1   Learning Platformer Level Probabilities

Now let's bring this concept back to content generation. Recall in the previous chapter we introduced a constraint-based n-gram approach for generating platformer levels, where we learned which vertical slices from a level are able to follow which other vertical slices. In this section, we will instead learn a probability distribution over the slices and use this to generate new level sections.

First, we need to define the random variables for this problem (in this case there is only one random variable); specifically, what are our possible outcomes? Fig. 6.1 shows a section of a platformer level that we will be using as our training data. Notice that each unique vertical slice is assigned a unique label. As we can see, some of these columns appear multiple times in the level, and our goal is to model the frequency of those unique columns with probabilities. We will treat each of the unique columns as one of our desired outcomes. Thus we can define our random variable as $X = \{A, B, C, D, E\}$, indicating the 5 unique columns.

To train our model we need to determine the probability of choosing each column, $P(X = \text{some column})$. We can estimate this distribution in a similar manner to how we estimated the distribution for the weighted die above. We can observe the frequency of the outcomes (different columns) in the training level, and construct a distribution for the

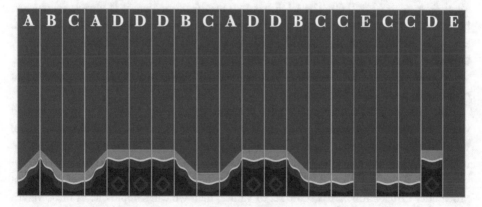

Fig. 6.1 This figure shows an example section of a platformer level where each unique vertical slice is indicated by a different letter

Table 6.2 This table shows how many times each unique column appears in the level (Occurrences), the probability of each unique column occurring ($P(X = $ Column$)$)

Column name	Occurrences	$P(X = $ Column$)$
Column A	3	$\frac{3}{20}$
Column B	3	$\frac{3}{20}$
Column C	6	$\frac{6}{20}$
Column D	6	$\frac{6}{20}$
Column E	2	$\frac{2}{20}$

outcomes using those frequencies. For example, column A occurs 3 times, and column C occurs 6 times. Table 6.2 shows how many times each unique column appears in the level in the *Occurrences* column; in the $P(X = $ Column$)$ column we see the probability of each unique column occurring, represented as the number of occurrences divided by the total number of columns in the level.

Now that we have the probability distribution, we can use it to generate new levels. To do this, we need to **sample** from the distribution until we have the number of columns we want. Sampling from a probability distribution simply means choosing a value for the random variable according to the likelihood of the values in the probability distribution. There are a number of ways to randomly sample values from a distribution, but we will use a straightforward and general approach called **inversion sampling**. Inversion sampling leverages the **cumulative distribution function (CDF)** of a probability distribution. The CDF of a probability distribution is the sum of probabilities for all values up to a given value and given some ordering of the variable values. For example, for the column distribution we just estimated, $CDF(A) = P(A) = \frac{3}{20} = 0.15$, and $CDF(C) = P(A) + P(B) + P(C) = \frac{3}{20} + \frac{3}{20} + \frac{6}{20} = \frac{12}{20} = 0.60$. The CDF has two useful properties: it is always increasing, because it is a cumulative function; and the values of the function always fall between 0 and 1. To perform inversion sampling with the CDF, we first randomly choose a value, r, uniformly from the range $[0,1]$. We can think of r as representing a probability value. Next, we apply the inverse CDF function to r, which tells us the value of the random variable that r corresponds to: $CDF^{-1}(r) = s$.

This sampling procedure may sound complex, but we are essentially trying to find where this randomly sampled value fits in the CDF. Figure 6.2 shows an example. Here, we randomly chose $r = 0.75$, and we can see that this value, when passed through the inverse CDF, gives Column D, because $CDF(D) = P(A) + P(B) + P(C) + P(D) = \frac{3}{20} + \frac{3}{20} + \frac{6}{20} + \frac{6}{20} = \frac{18}{20} = 0.90$, and $CDF(C) < r \leq CDF(D)$. Practically, if we have a probability distribution across some set of values, then we can sample one of these values by randomly choosing a probability value, r, and seeing where that value falls in the dis-

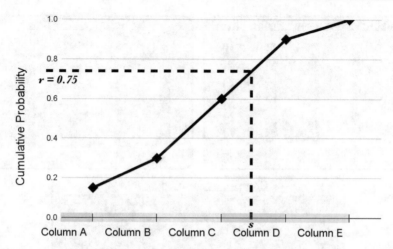

Fig. 6.2 This figure shows a plot of the cumulative distribution function (CDF), and an example of performing inversion sampling with $r = 0.75$

tribution (e.g., by adding up individual probabilities in an ordered way until you find the variable value that makes the sum greater than r).

Now let's get back to our level generation example. In our case there are 20 total observations so we can imagine using a 20-sided die to choose an r value to perform inversion sampling. For example, rolling a 2 would give $r = \frac{2}{20} = 0.1$. If we look at Fig. 6.2 again, we can see that $r = 0.1$ selects column A. Similarly, rolling an 8 would give $r = \frac{8}{20} = 0.4$, which would select column C. We are using a 20-sided die in our example for clarity and ease of understanding, but in digital PCG systems a standard random number generator would be used. Figure 6.3 shows the process of sampling a new level column by column using a 20-sided die. The top-left of the figure shows all the possible column values to choose from (and their associated probabilities), while the top-right shows how rolling the die relates to the choice of column. In the figure we can see how after each die roll a new column is chosen, increasing the length of the level. In the bottom of the figure we can see a new level of the same length as the input section generated by our model.

In the complete generated level the distribution of some of the columns is similar to the distribution of columns in the training level. Specifically, we can see that column C's frequency is the same as in the training level, 6; and column D's is very close as well, 5 in the generated level and 6 in the training level. However, some of the columns are less similar to their observed frequencies. Column A appears 7 times in the generated level, but only 3 times in the training level! This is a result of the random sampling performed. If we were to continue sampling this level for another 50, 100, 1000 columns eventually the distribution of columns in the generated level would approach the distribution in the training section. But if we compare the sampled and training level we can see that the arrangement of the columns is quite different. This can be a desirable quality in the output if any relative

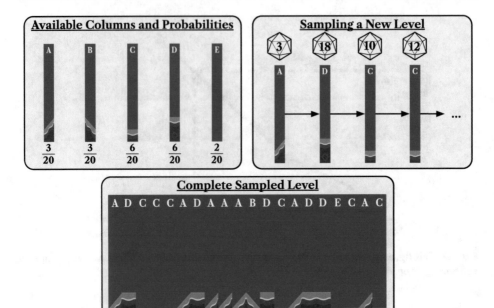

Fig. 6.3 This figure shows the possible columns (top left), how sampling could be performed using a 20-sided (top-right), and the resulting sampled level (bottom)

ordering of the columns is acceptable. Unfortunately, looking at the generated level, there are sequences that appear somewhat disjointed (e.g., multiple *A* columns in sequence leads to a jagged look). Furthermore, a poor ordering of columns could lead to impassable areas (e.g., if column *E* is chosen many times in a row). Observing these issues, we can get an intuition that the next column to place might depend on which columns were placed before it. More concretely, we can start thinking of the placed columns not as independent random events, but instead reliant on the **conditional probability distribution** of the columns.

6.2   What are Conditional Probabilities?

Above we introduced basic probabilities and an example of how you can use them for generating content. But what if we are interested in more than just raw frequency? In that case we need to capture more context around the patterns and their occurrences. One way to capture more context in the model is to use conditional probabilities. Conditional probabilities capture how likely an event is to occur given some other event has already occurred (or some other relevant information is known). Formally, this can be written as the probability of event *A* happening, given that event *B* has already occurred, or $P(A|B)$. This conditional

Table 6.3 This table shows all possible dice combinations. Combinations where the total is 8 or more are **bold**, and combinations where the first die was 4 are *italic*

	1	2	3	4	5	6
1	2	3	4	5	6	7
2	3	4	5	6	7	**8**
3	4	5	6	7	**8**	**9**
4	*5*	*6*	*7*	*8*	*9*	*10*
5	6	7	**8**	**9**	**10**	**11**
6	7	**8**	**9**	**10**	**11**	**12**

probability is equal to the probability of both events A and B occurring, divided by the probability that event B occurs:

$$P(A|B) = \frac{P(A \cap B)}{P(B)} \tag{6.1}$$

To discuss more concretely, let's return to the dice rolling example from before. Say we have someone roll a fair die, and we want to know the probability that the total after rolling another die will be greater than some value. Let's imagine we're playing a table top RPG where you and a partner need to roll a combined value over a certain amount to land a combination attack on an enemy; your partner has rolled, and you now get to decide your action based on their result (i.e., you can roll your die to try to land the combination attack, or decide it's too risky and just shield yourself). Then we would be trying to capture the probability that the total will be greater than a certain value given the value of the first roll, a perfect opportunity to use conditional probabilities. For example, if your partner rolled a 4, and we need at least 8 total to hit the enemy, then what we want to compute is $P(D_1 + D_2 \geq 8 | D_1 = 4)$, where D_1 is the first die and D_2 is the second, unrolled die. This would give the probability of successfully hitting the enemy, which could influence how you choose to proceed in the game. If we visualize the set of all possible dice combinations as a table, then we can compute these values more easily.

Now, we first need to compute $P(D_1 = 4)$, or the probability of rolling a 4. Since it is a fair, six-sided die we can assume this is $\frac{1}{6}$. Next, we need the probability of rolling a 4 **and** rolling a total of at least 8. We can do this by checking the combinations of rolls that result in at least 8, and then look at how many of those include a 4 for the first die. Table 6.3 shows all the possible dice combinations, and any combination where the total is 8 or more is bolded. This takes care of one part of it. However, we are only interested in the results where both the sum is at least 8 **and** we roll a 4 first. To show this second part, we italicize all combinations in Table 6.3 where the first die is rolled as a 4. From here, we just need to count the number of bold and italic entries in the table indicating where both statements are true (i.e., first die is 4, total is at least 8). Counting these up gives us 3, which gives us the probability $\frac{3}{36}$, since there are 36 possible dice combinations. Finally, we can use the equation defined before to get the resulting conditional probability:

$$P(D_1 + D_2 \geq 8 | D_1 = 4) = \frac{P(D_1 + D_2 \geq 8 \cap D_1 = 4)}{P(D_1 = 4)} = \frac{\frac{3}{36}}{\frac{1}{6}} = \frac{18}{36} = \frac{1}{2}. \qquad (6.2)$$

So if you roll a 4, there is 50% chance you roll at least an 8 total with the second die. We can also see this in the table; there are 6 total italicized cells indicating outcomes where a 4 is rolled first, and half of them are bold, indicating a sum of at least 8.

There is another way to compute and represent this scenario that maps more directly to how such a model might be learned for content generation. Here we will instead directly model the probability of rolling an 8 or greater given any first die roll. First we need to define our random variable (i.e., the possible outcomes). For this example we will simply assume two outcomes, (1) the sum of the two rolled dice is at least 8 or (2) the sum of the two rolled dice is less than 8. We could also treat each possible sum as different outcomes, but this representation will simplify the example for now. Next, we need to define the possible events on which to condition the probability (i.e., the B in $P(A|B)$). Since we are interested in the sum given the value of the first die's value, we will treat the value for the first die roll as the condition. We can then represent the conditional probability distribution using a table like in Table 6.4, where the first column indicates the condition (i.e., the possible first die value), and the other columns corresponds to the possible outcomes (i.e., less than 8, or at least 8). With this table layout, each row corresponds to the probability distribution for a different condition. We fill in the table by checking each condition, and counting up the ways that each outcome could occur, and dividing those values by the total number of different ways each outcome could occur for that condition. For example, if we think about starting with a roll of a 4 on the first die, then we can count up the ways we could roll a total less than 8 (i.e., rolling a 1, 2, or 3 on the second die) and the ways we could roll at least an 8 (i.e., rolling a 4, 5, or 6 on the second die). These counts then give us our conditional probabilities seen in Table 6.4.

6.2.1   Learning Platformer Level Conditional Probabilities

Now let's bring conditional probabilities to content generation, by looping back to the platformer level generation problem we were discussing earlier. Previously, we were able to generate a new level by sampling from the distribution of level columns. Here we will be using the same representation, but instead of learning just the distribution of those columns, we will be learning conditional probability distributions for the columns occurring after one another.

So again we return to the platformer level in Fig. 6.1 to train our model. We still have the same set of columns that we observed previously (A-E), but now we are building a model where the next placed column depends on those already placed. Notice that this model is

Table 6.4 This table shows the conditional probabilities for rolling either less than 8 (center column) or at least an 8 (right column), given the value of the first die

First die	Less than 8	At least 8
1	$\frac{6}{6}$	$\frac{0}{6}$
2	$\frac{5}{6}$	$\frac{1}{6}$
3	$\frac{4}{6}$	$\frac{2}{6}$
4	$\frac{3}{6}$	$\frac{3}{6}$
5	$\frac{2}{6}$	$\frac{4}{6}$
6	$\frac{1}{6}$	$\frac{5}{6}$

Table 6.5 This table shows the conditional probabilities for placing each column type (i.e., A, B, C, D, E), given the value of the previous column

Previous column (Col_{t-1})	$P(Col_t|Col_{t-1})$				
	A	B	C	D	E
Column A	$\frac{0}{3}$	$\frac{1}{3}$	$\frac{0}{3}$	$\frac{2}{3}$	$\frac{0}{3}$
Column B	$\frac{0}{3}$	$\frac{0}{3}$	$\frac{3}{3}$	$\frac{0}{3}$	$\frac{0}{3}$
Column C	$\frac{2}{6}$	$\frac{0}{6}$	$\frac{2}{6}$	$\frac{1}{6}$	$\frac{1}{6}$
Column D	$\frac{0}{6}$	$\frac{2}{6}$	$\frac{0}{6}$	$\frac{3}{6}$	$\frac{1}{6}$
Column E	$\frac{1}{2}$	$\frac{0}{2}$	$\frac{1}{2}$	$\frac{0}{2}$	$\frac{0}{2}$

a probabilistic variation of the constraint-based n-gram model from Sect. 5.1. To train this model we will again define our conditions. In this example we will use $n = 2$ (indicating pairs of columns), and use only the immediately previous column as the condition. Therefore, to construct the conditional distribution, we count how many times each column follows each other column. This gives us Table 6.5, where the leftmost column of the table indicates the condition (i.e., the previous column, Col_{t-1}), and the other columns in the table indicate the probability of each column type following the previous column type (i.e., $P(Col_t|Col_{t-1})$). For example, in row D we can see that there is a 2/6 chance of a B column following a D column, and a 3/6 chance of another D column. Notice that now instead of dividing by the total columns in the level, we are instead dividing only by the total number of columns for that given condition (e.g., 3 observations with Column A as the condition, and 6 observations with Column C as the condition). With this trained model, we can now try to generate a new level.

To generate a new level, we need to choose a starting column. If we have more examples levels, we can model this directly by adding an artificial column to the start of each level.

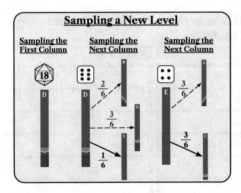

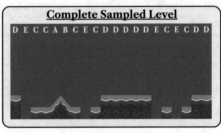

Fig. 6.4 This figure shows how to generate a new level column-by-column by sampling from the learned conditional probabilities using a 6-sided die (left), and the resulting sampled level (right). Note, during sampling the bolded arrow indicates the selected column

Then during training, we can directly learn which other columns follow that artificial starting column. However, since we only have one example level, let's choose the starting column by sampling from the independent distribution of the columns (trained in Sect. 6.1.1).

Figure 6.4 shows how we can generate a new level using our trained conditional distribution. The left shows how we can choose the first column using the unconditional probabilities learned earlier, and then how each subsequent column is chosen by sampling from the appropriate conditional distribution. We can sample from this distribution using much the same method as before. The only difference here is that we need to use the correct condition for our distribution to ensure we are referencing the correct distribution. For example, we chose column D for the first column. So to choose the next column, we consult row D of the table and randomly choose the next column according to the distribution there, again using the inversion sampling approach. For this example, we can use a 6-sided die to simulate the random choice of r for inversion sampling, since (luckily, or by design) all the conditional observation totals are factors of 6.

We can see that columns B, D, and E have a $\frac{2}{6}$, $\frac{3}{6}$, and $\frac{1}{6}$ probability of following a D column, respectively. And so we can roll a 6-sided die to determine which outcome to choose. We roll a 6, so we place an E column, and repeat the process. The right of Fig. 6.4 shows the resulting generated level. Notice, as compared to the level generated in Sect. 6.1.1 with the unconditional probabilistic model, this level does not have disjointed sequences. That is because using the conditional distribution trained in this way ensures that all sequences of 2 columns that appear in the output have appeared in the input. The benefit being, as stated, that there are fewer strange layouts occurring in the output. A drawback of this approach, however, is less variety in the output (particularly with a small training set). In this case, during training there are some column types that only have one possible following column, which can result in overly repeated patterns. For example, if column B is sampled, then column C must be the next column according to the trained distribution.

On the reverse side, there are also sequences of columns that will never occur with this distribution, but which may be desirable. For example, column E following column A could be a reasonable sequence. One way of combating both of these issues to introduce some smoothing to the distribution. One such approach is additive or Laplace smoothing [113], where you essentially add a small number to each possibility in the distribution during the estimation stage (i.e., when counting how frequently each outcome happens). More formally, if we use a smoothing factor of α, then instead of directly using

$$P(Col_t|Col_{t-1}) = \frac{\text{Count}(Col_t \text{ following } Col_{t-1})}{\text{Count}(Col_{t-1})} \tag{6.3}$$

we would use

$$P(Col_t|Col_{t-1}) = \frac{\text{Count}(Col_t \text{ following } Col_{t-1}) + \alpha}{\text{Count}(Col_{t-1}) + (\alpha \times \# \text{ of Column Types})}. \tag{6.4}$$

For a more concrete example, if we take the conditional distribution for column B, and use $\alpha = 1$, then we get:

$$\begin{array}{cccccc} & P(A|B) & P(B|B) & P(C|B) & P(D|B) & P(E|B) \\ B & \frac{0+1}{3+5} = \frac{1}{8} & \frac{0+1}{3+5} = \frac{1}{8} & \frac{3+1}{3+5} = \frac{4}{8} & \frac{0+1}{3+5} = \frac{1}{8} & \frac{0+1}{3+5} = \frac{1}{8} \end{array}$$

This ensures that each event has a non-zero chance of occurring, which (1) increases the possibility space, (2) decreases the certainty of outcomes in areas of the distribution where little data exists, and (3) does not have a large effect on the areas of the distribution that are well observed. As a note, for domains where the number of possibilities is very large this form of smoothing can distort the probability distribution to be quite different from the observed probabilities. For those cases there exist other more complex forms of smoothing that can avoid such problems (e.g., Kneser-Ney smoothing [131]). When we do not use any smoothing, then we will get results similar to the constraint-based n-gram approach from Chap. 5; we won't produce any patterns that weren't observed during training. The key difference between the constraint-based and probabilistic versions of the n-gram approach is that the more common patterns from the training data will appear more frequently when using the probabilistic version. This might not be noticeable when comparing individual outputs, but if we generated many levels it would be more noticeable.

We have introduced the basics around probabilities, conditional probabilities, and how you might model and generate content with these distributions. Next we walk through different PCGML approaches that rely on these types of distributions.

6.3   Markov Models

The first set of approaches we'll be discussing are **Markov** models [120, 121]. Markov models are a set of probabilistic models that rely on the assumption that the next state or event only relies on the current state or event, ignoring the history of how you got to that state. This assumption is referred to as the **Markov property**. The Markov property is useful when modeling complex domains, as it allows the simplification of the problem from modeling "what is the relationship between all states" to "what is the relationship between the current and next states." Markov models are a popular family of probabilistic models, and have been used frequently in the context of PCG [21, 181, 218]. There are a number of different types of Markov models, and in the following sections we'll discuss some that have been used for content generation.

6.3.1   Markov Chains

The simplest Markov model is the **Markov chain**. A Markov chain models the probability of transitioning from one state to another, without considering the states before the current one. Formally, a Markov chain is defined as a set of states $S = \{s_1, s_2, ..., s_n\}$ and transition probabilities $P(S_t|S_{t-1})$, i.e., the probability of transitioning to a state ($S_t \in S$), given the current state ($S_{t-1} \in S$). Since Markov chains model transitions between states, they can be visualized as state machines where the edges are the transition probabilities and the nodes are the different states. If we look back at the model we trained in Sect. 6.2.1, we can see that it satisfies the Markov property. In fact, this model is an n-gram, and specifically a bigram, which models the relationship between pairs of columns; or in terms of a Markov chain, it captures the probability of transitioning from the current column to next column. Here we can represent this bigram model as a Markov chain, where the unique columns make up the set of states; and the conditional probability distribution (Table 6.5) gives the transition probabilities. Figure 6.5 shows a state machine representation of the trained model, where nodes correspond to the unique columns and arrows indicate that the following column was observed after the current column with the indicated probability. With this representation, we can more easily visualize the state transitions that are taking place when sampling a new level. Each time a new column is chosen, it is the same as following an edge from one node in the graph to another.

This model only captures the relationships between pairs of columns, which can make it difficult to capture larger structures and patterns (e.g., the plateau pattern of columns $ADDB$), particularly if more training data is provided. Luckily, Markov chains can be extended to account for not just the immediately previous state, but a finite number of previous states. Continuing with this n-gram approach, imagine if instead of pairs of columns we were looking at triplets? That is, what if the next column depended on the previous 2 columns? Then this is called a trigram, and we could train the conditional distribution in much

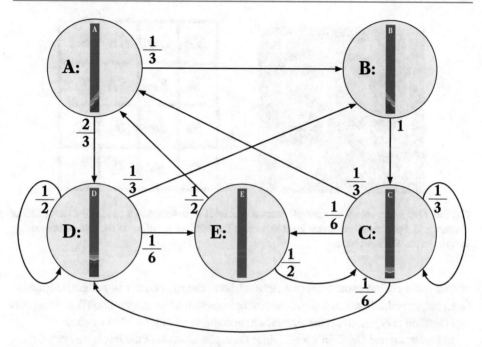

Fig. 6.5 This figure shows how we could represent the *n*-gram model learned in Sect. 6.2.1 as a state machine. Each node represents a unique column, and each arrow represents that the corresponding column is able to follow the current column with the indicated probability. By using the learned conditional probabilities as state transition probabilities, we can more easily think of this model as a Markov chain

the same way as before. The difference here is that instead of learning $P(Col_t|Col_{t-1})$ we will be learning $P(Col_t|Col_{t-1}, Col_{t-2})$. Extending the Markov chain in this way is called a Markov chain with memory, or an **order n Markov chain**, in this case, an order 2 Markov chain because we are capturing 2 previous states. Extending the memory or raising the order of the Markov chain allows the models to capture more context during training and sampling, and can lead to more cohesive output with more consistent replication of larger patterns from the training data.

6.3.2 Multi-dimensional Markov Chains

In addition to the *n*-gram approach discussed above, and the related research that has explored their use [33], order *n* Markov chains can be used with other level representations, resulting in very different models. A common representation of a game level is the tile-based representation which we mentioned in Chap. 5. A tile-based representation splits a level into a 2 or 3-dimensional grid, where each cell in the grid is labeled using a set of tile types

Fig. 6.6 This figure shows an example section of a level represented in a tile format (left), and an abstract grid representation for that level indicating the different positions in the grid corresponding to states in the Markov chain

corresponding to level structures (e.g., ground, item, enemy) or other important information (e.g., player path). Figure 6.6 shows an example section of level split into tiles. Snodgrass and Ontañón [177] used this representation for training an order n Markov chain.

In Snodgrass and Ontañón's work, since the representation of the levels was two dimensional, the Markov chain also needed to be able to handle dependencies in two dimensions. To address this, they re-represented an order 3 Markov chain to work on a 2D grid, calling it a **Multi-dimensional Markov Chain** or **MdMC**. This MdMC approach works in a manner very similar to an order n Markov chain, with the main difference being the structure of how the previous states are defined. In this multi-dimensional space, there are many ways to capture state dependencies and represent them in the model. We could, for example, treat each row as its own generated sequence. Or we could have the end of one row feed into the start of the next row. Or we could do a snaking pattern where we are reversing the order of the row generation each time we reach the end (i.e., generate $S_1 \rightarrow S_4$, then $S_8 \rightarrow S_5$, with S_4 feeding into S_8). We could also generate column by column instead of row by row and use those same options. For simplicity, let's assume we want to generate row by row, and we want to just go left to right with the last element of a row not feeding into the next row.

If we were to train a standard Markov chain (order 1) on a 2D level in this way, then each state in a row would only depend on the previous state in that row, and the states in different rows would not affect one another. Figure 6.7 (left) shows the dependencies between different positions in the example tile level section; notice that none of the positions depend on any positions in the previous rows. This setup would result in every row being generated independently, which could create some strange structures (e.g., a ground tile floating in the middle and top of the level). Additionally, we can train an order 2 Markov chain, but still only focusing on one dimension. Figure 6.7 (middle) shows the dependencies between different positions in the example tile level section with an order 2 Markov chain. Notice again that the rows do not have interactions between each other, and so this setup would have the same

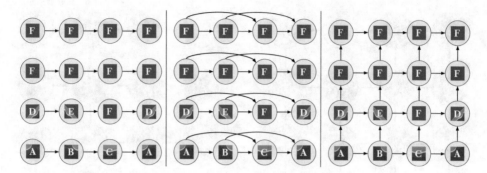

Fig. 6.7 This figure shows potential state and positional dependencies for Markov chains in 2D domains. The left shows a standard Markov chain in this domain. The middle shows a simple order 2 Markov chain. And the right shows an order 2 Multi-dimensional Markov chain where dependencies between the rows are captured

issues as the order 1 Markov chain above. Lastly, we can take the MdMC approach, and reconsider what is included as a previous state in these 2D levels. Figure 6.7 (right) shows how an order 2 Markov chain's dependencies can be reordered to account for the positions in two dimensions. Here each position uses the previous position in the horizontal and vertical axes as its context (i.e., its previous states). This setup can yield a model that better captures the patterns present in 2D (or 3D) domains. Snodgrass and Ontañón [177] used this approach (and variations of it) to model and generate levels for 2D platforming games.

Multi-dimensional Markov chains work well in domains where there is a strong concept of sequencing. That is, domains where it makes sense to assume that position x is dependent upon position $x - 1$. In terms of game levels, this typically means that levels are meant to be traversed in a particular direction (e.g., in *Mario* games you typically move to the right). What if we are instead interested in domains without a strong sense of directionality (e.g., a game like *Zelda* where you need to move in all directions)? In these cases, we can rely on a different set of Markov models.

6.3.3 Markov Random Fields

Recall that Markov chains model dependencies in a sequence of states or random variables, where the next state depends on a subset of previous states. Markov random fields (MRFs) do not use this concept of "previous states" or sequences and instead work on an undirected graph of states, where each node in the the graph is a random variable. MRFs satisfy the Markov property by only depending on states within their **Markov blanket**. Formally, a Markov blanket for a random variable S_i in a set of random variables $\mathbb{S} = \{S_1, ..., S_n\}$, which is defined as any subset, $\mathbb{S}_j \in \mathbb{S}$, of random variables from $\mathbb{S}$ such that $P(S_i|\mathbb{S}_j) = P(S_i|\mathbb{S})$. More intuitively, this means that conditioning a node on its neighbors (defined by the Markov

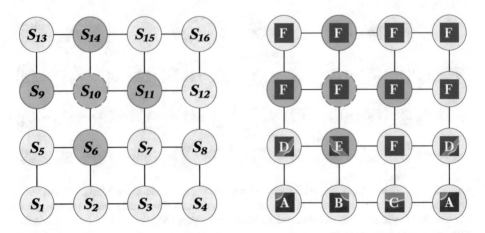

Fig. 6.8 This figure shows the undirected graph representing an abstract level (left). The dotted blue state (S_{10}) has its Markov blanket (neighbors) highlighted in blue. The right show how our example level aligns with this model

blanket) gives all the needed context for that variable, and adding additional nodes for context will not change how that node or variable is sampled.

Figure 6.8 (left) shows how an abstract grid level could be modeled with an MRF; and what that looks like for our example level (right). An MRF is formally defined by (1) an undirected graph $G = (V, E)$, where V is a set of vertices and E is a set of edges connecting those nodes (which defines the Markov blanket and conditionals); and (2) a set of random variables S, where each $S_i \in S$ represents the label or value for a given vertex $v \in V$. Note that one of the main advantages of using an MRF over a Markov chain approach is that an MRF is able to capture a wider range of dependencies in all directions without relying on sequential structures. For example, let's say we have a pattern where a treasure tile is surrounded by trap tiles. A Markov chain approach would try to capture this sequentially using contextual information that captured parts of the pattern (i.e., the treasure follows the traps, and then traps follow the treasure). But an MRF approach can more easily capture the treasure surrounded by traps patterns by observing the tiles surrounding the treasure instead of just the tiles preceding the treasure.

Now, even though MRFs rely on different structures and dependencies, the training procedure follows much the same process as the Markov chain approaches (and n-gram approaches) above. Namely, we can simply count up how frequently each value occurs with each different surrounding configuration. Essentially, when going through the examples instead of looking at the previous column or tiles, we instead look at the neighboring tiles in the graph to define the condition for the distribution. The differences arise when using the trained MRF models to sample new content. Since MRFs do not have this concept of directionality, it may not be obvious how to sample from them. For sampling approaches we can look to the domain of image and texture synthesis where MRFs have been used frequently for years [140].

There are a few different ways we can leverage MRFs for generation. First, we can do what is called **infilling**. Infilling is a process by which you start with a partially completed image (or texture, level, etc.) and use the MRF to fill in the remaining details. Infilling works by choosing the position in the graph with the largest number of its neighbors labeled, while still being unlabeled itself, and then setting the label of that position according to the trained conditional distribution. Setting the label can be done by either choosing the most likely option or by sampling from the distribution. This way of generating content from an MRF is used primarily in texture synthesis and image generation, or as a way of evaluating how well a model captures a domain (e.g., can it reproduce the training or test data). As of writing, infilling with MRFs has not been used for level generation or other content generation for games. However, infilling could be an interesting way of enabling some co-creative tools leveraging MRFs. For example, we can set up a system where a user partially specifies the structure of a level and then has the MRF fill in the rest of the unspecified space. In Chap. 11 we discuss mixed-initiative PCGML tools in more detail.

The other approach for generating with MRFs is to start with an initial (randomized) level, and find the best arrangement of tiles in that level (in terms of likelihood with the trained model). In practice, comparing all the possible tile arrangements for a level to find the optimal arrangement is impractical. Even our very small 4×4 level example in Fig. 6.8 (right) with a lot of repeated tile types has over 14 million possible arrangements (i.e., $\frac{16!}{9!2!2!1!1!} = 14414400$), and this number explodes as the size of the levels increases and as the number of tile types increases. So instead we can start with a randomized level and iteratively improve the arrangement of tiles using a **Markov Chain Monte Carlo** sampling approach (**MCMC**) [4]. MCMC sampling combines Monte Carlo estimation and Markov chain sampling. Monte Carlo approaches use randomness to produce answers that approach the true answer as the numbers of samples increases, and are commonly used for problems that are impossible (or very hard) to solve deterministically. Markov chain sampling (as discussed previously) is sampling a new value or state using only the current state as context. In the case of Markov Chain Monte Carlo, we have a distribution that we want to sample from, but directly sampling from it is impossible due to the number of possibilities. So instead, we construct a Markov Chain whose distribution will tend to the distribution that we want, and then we can take samples from it to get our desired distribution. Abstractly, this can sound a bit daunting. In our case we are trying to estimate the optimal arrangement of tiles in a level given our trained distribution (the Monte Carlo part). We do this by making incremental sequential changes to the arrangement of the level (the Markov chain part). Van Ravenzwaaij et al. [215] provide a more detailed but easy to follow introduction to MCMC.

For MRF sampling, we can use the **Metropolis-Hastings** sampling algorithm [70], which is one such example of an MCMC sampling approach. This sampling algorithm requires a proposal method and an acceptance method. That is, a way of suggesting a change to the level layout, and way of deciding if that change should be kept or discarded. Here we will describe the proposal and acceptance methods presented alongside the algorithm [70]. The proposal method simply chooses two positions randomly and swaps the tiles at those positions. The acceptance method, computes the pre-swap and post-swap likelihoods for the level, given the trained MRF distribution. This method will then accept the swap if it improves the likelihood, and reject the swap with probability proportional to the decrease in likelihood. To actually sample a new level with this approach, we start with a fully labeled graph, where the labels are assigned randomly according to the frequency of labels in the training data (our initial state). Next, we do an iteration of our proposal and acceptance methods. Specifically, we choose 2 positions in the graph, swap the labels, and decide if we should keep the swap using the likelihoods. One iteration of proposing and accepting/rejecting gives us the next state of our Markov chain in our MCMC sampling. This label swapping procedure is repeated until the overall likelihood of the graph labeling stops improving or some other defined stopping criterion is reached (e.g., a maximum number of iterations). This stopping criteria is the Monte Carlo approximation part; we assume with the stopping criteria that we have approximated the best arrangement of the level for the distribution. This Metropolis-Hastings sampling approach has been used by Snodgrass and Ontañón [181] when generating platformer levels as well as by Volz et al. [218] when generating match-3 puzzle boards.

The Markov models and examples we've discussed so far have all directly modeled the tiles in the level. But what if there is additional information outside of the tile structure that could give our models more context about the patterns and structures to learn? For example, what if we are modeling building structures from *Minecraft*, and we want to indicate what type of building we are modeling (e.g., a cottage vs. a castle)? In these cases, we can try using **conditional random fields**.

Conditional Random Fields

An interesting extension to Markov random fields are **conditional random fields** (**CRFs**). CRFs are defined similarly to MRFs as a graph $G = (V, E)$, and a way of labeling the vertices, X. The difference is that the label of each vertex is also globally conditioned on some other value. This can be likened to having a separate MRF for each possible condition in the extra variable of the CRF, or simply adding another condition to the conditional distribution. If we are modeling textures, this extra variable could indicate if the texture is natural (e.g., wood, stone, etc.) or synthetic (e.g., spirals, checkerboard). CRFs have the added benefit of being able to provide further context to the model. This additional context does not even need to represent the same information captured by the previous MRF labels. For a more concrete example, if we are modeling a level using an MRF then all the context the MRF gets for each position is the labels of that position's neighbors (i.e., $P(S_{(x,y)}|neighbors(S_{(x,y)}))$. If instead we are using a CRF to model the level then we can still condition on the neighbors, but we can also globally condition all the positions on a label relating to meta information about the level overall (e.g., difficulty, level type, etc.) or to some other relevant information (e.g., current height in the level, $P(S_{(x,y)}|neighbors(S_{(x,y)}), height = h))$. Adding this additional context can improve the reliability of the patterns modeled and enable more control over the model during generation. However, additional context does require additional training data to fill out the distributions. As of writing, CRFs have not been used for PCGML, but similar ideas of additional external labels and context have been explored for MdMC models [182].

6.3.4 Other Markov Models

So far we have focused on the Markov models that have been used in the context of PCGML. There are more Markov models that we will not discuss in detail, but will instead mention here as potential avenues of future PCGML approaches. One such model is a Markov Decision Process (MDP), which is a way of modeling sequential decisions. We leave the detailed discussion of MDPs to Chap. 10 where we introduce how they relate to reinforcement learning. Other such models are Hidden Markov models (HMMs) [141] and Hidden Markov random fields (HMRFs) [100]. The *hidden* part of these models comes from the assumption that we cannot actually observe the states of the system, but only the output of the states. Furthermore, the output of those states is stochastic. For comparison, in a standard Markov chain each state has only one output, and so we can equate the state and the observed output (e.g., in our level generation example from Sect. 6.1.1, the state corresponds to the column type). In a hidden Markov model, the states output different values based on an internal distribution. Again, in our level generation example, there could be a state that outputs column C or column D with equal probability. These hidden states are often mapped to some semantic domain meaning. In the level generation example, there could be different states corresponding to types of level sections or structures (e.g., enemy-heavy section, jumping-heavy section, relaxing section, etc.). If instead we are modeling charts for a rhythm game, the hidden variable could indicate the intensity of a section (i.e., roughly how difficult the section is). This formulation essentially gives another contextual variable to help explain the observed values in the data. With HMMs and HMRFs, the number of hidden states is typically selected; and so training a model usually boils down to learning the hidden state *transition* probabilities and the hidden state *emission* probabilities. That is, how can we get from one hidden state to another, and when in a given state what can we expect to observe.

6.4 Bayesian Networks

Above we discussed various Markov models, and saw how they can be used to model sequences and undirected graphs. But what if we want to model graphs with directional dependencies? Well, we can't use MRFs since they are by definition undirected. We could try to formulate the dependencies in such a way that Markov chains could be used, but that could lead to cycles in the graph, which might not be desirable. Instead, we could turn to Bayesian networks. Bayesian networks (BNs) are a probabilistic graphical model,

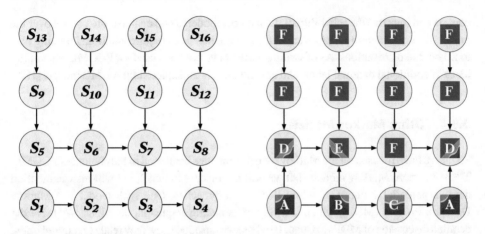

Fig. 6.9 An example directed acyclic graph representing the abstract level (left). This graph can be used as a Bayesian network where each node represents a position in the level and edges indicate a one-way dependency between two positions. The right shows how this graph maps to our example level section

but unlike MRFs, BNs leverage directed acyclic graphs (DAGs). This means that BNs are able to capture one-way dependencies (directed edges), but are not able to capture cyclic dependencies. If we return to the example of modeling a 2D level, then in the simplest formulation for a BN model, we would manually define the network structure as a DAG using our knowledge and assumptions about the structure of the domain. Figure 6.9 shows one way we might formulate the dependencies between positions in a 2D level.

Using a predefined network, training and generation are similar to Markov chains discussed above. Namely, the conditional distribution for each node could be estimated from the observations in the training data. Similarly, generating a new level with a predefined network can follow the same process of first assigning the values to the unconditioned nodes (in the case of Fig. 6.9, nodes S_1, S_{13}, S_{14}, S_{15}, S_{16}), and then the other nodes can have their labels assigned following some ordering.

If we are not domain experts or the domain is complex, then we might not want to manually define the structure of the network. In this case, we can try to learn the structure of the network from the data itself. There is a large body of research on this topic [12], and covering it in detail is out of scope for this book. But we will briefly highlight the two families of approaches: score-based and constraint-based. Score-based approaches work by first defining some metric by which to evaluate the quality of a network, and then searching the space of possible structures to find the best (or good enough) structure. Often the metric used is related to the likelihood of the data under the trained model and structure. Constraint-based approaches try to identify a set of independence constraints between variables/nodes from the data, and then search for a graph structure that satisfies those constraints.

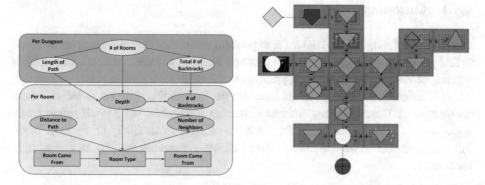

Fig. 6.10 The Bayesian network used by Summerville et al. [193, 197] for modeling and generating dungeon maps (left), and an example annotated dungeon map where each symbol indicates a different room feature or type. (recreated with permission)

Bayesian Networks have been used to generate levels [59, 197] as well as full games [63]. One interesting example of using a Bayesian network for level generation comes from Summerville and Matteas [193]. In their work they model and generate *Legend of Zelda* dungeon layouts, specifically the placement and connections of rooms in the dungeons. Figure 6.10 shows the network used by their model (left), and how they annotated their example dungeons to indicate different types of rooms (right). Notice that their network depends not only on structural level data (e.g., room type, number of neighbors), but also play information (e.g., path length). The other approaches mentioned above rely on topics that will be introduced in the following section, and so we will discuss them in more detail shortly.

6.5   Latent Variables

When discussing Hidden Markov Models, we mentioned that they relied on additional hidden variables that we couldn't directly observe. These hidden variables are more commonly called **latent variables**, and they occur frequently in complex domains and systems. Latent variables will play a big role in Chaps. 7, 8, and 9 when we begin discussing neural networks, but for now we will just introduce some of the simple ways to model latent variables using the observed data. In HMMs, the latent variables are defined a priori (i.e., we set how many hidden states there are), and during training we estimate the transition and emission probabilities of the hidden states from the data as mentioned in Sect. 6.3.4. But what about a case where you do not know what the latent variables are?

6.5.1   Clustering

One way of defining latent variables is to perform **clustering** over the examples available
to find groupings of the data, and using those group labels as latent variables. Clustering is
a family of unsupervised learning approaches that aims to separate data points into groups,
where individuals within a group are more similar to each than they are to individuals in
another group. There are many different clustering approaches, but they all aim to create
a grouping of provided data according to some measure of similarity or distance. In the
following subsections we will discuss some PCGML approaches that rely on clustering in
some form.

Hierarchical Multi-dimensional Markov Chains
One example of using a PCGML approach that relies on clustering comes from Snodgrass and Ontañón [178]. Figure 6.11 shows the pipeline they used for their approach. In
this approach they extended their previous multi-dimensional Markov chain (MdMC) by
representing levels at two resolutions: the standard tile-based resolution; and a high-level
resolution where the labels were defined through clustering (Fig. 6.11 (2, 3)). This approach
tried to define latent variables related to common structures and patterns in the training levels
by clustering chunks of tiles (e.g., 6 × 6) from the levels. Clustering these chunks led to
groupings of chunks with similar attributes, such as structural outlines or tile frequencies.
Using these identified clusters, they re-represented the training levels into an abstracted
tile-based representation. Then, one MdMC was trained on the levels in this abstracted representation, and was used to generate new levels in this representation. Another MdMC
was trained for each of the clusters to try to directly model the patterns common in each
cluster. This essentially led to one MdMC model for the high-level abstracted representation
and one combined model for the low-level tile representation, where the low-level model
is conditioned on which high-level section it is currently generating. This extra condition
can be seen as similar to the extra condition used in CRFs. Notice that this type of clustering to find common structures is something that can be added to a lot of PCGML models

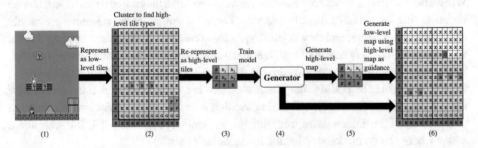

Fig. 6.11 The pipeline used by Snodgrass and Ontañón [178] for identifying common level structures
using clustering, and then generating new levels at two resolutions. (Recreated with permission)

(e.g., clustering level sections to train separate models per section type, clustering quest templates to identify more general patterns, clustering sprites to find common relationships with gameplay, etc.).

Bayesian Networks with Latent Variables

Another existing PCGML approach that relies on clustering is the Bayesian network approach by Guzdial and Riedl [59] that we briefly mentioned in Sect. 6.4. Figure 6.12 shows the different stages of this approach. This work represents sections of platformer levels as their sprite counts and a measure of how long the player has spent inside that chunk. They then perform two levels of clustering. First, they cluster the sections using the difference between the sprite frequencies and the interaction times to measure similarity between chunks. Next, they compute an additional feature indicating the relevance score for each sprite in its given cluster. A second clustering uses the difference between these relevance scores, the interaction times, and the placement within the level to get a final set of representative level sections. Since the resulting clusters are meant to capture the different groups of level sections based on structure and gameplay, separate BNs are then defined and trained for each of the identified level chunk clusters.

The BNs trained at this stage have a mix of observable nodes (for level geometry, number of sprites, and relative positioning of the sprite groups), and hidden nodes. The hidden nodes correspond to sprite shape "style," which is identified through clustering observable nodes; level section "style," which is identified through clustering the previous hidden node and the node for sprite numbers. In the end, this approach truly highlights the power of clustering by clustering at nearly every stage of training both for segmenting the training data and for directly defining latent variables in the models. Using this model to evaluate levels was even found to correlate with human evaluations of level style and design [60].

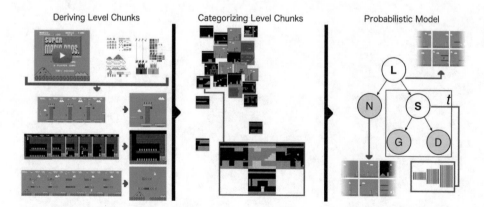

Fig. 6.12 Guzdial and Riedl's Bayesian network approach [59] extracts level sections (left), clusters them to find groups (center), and samples new sections (right). (Recreated with permission)

6.6   Takeaways

Probabilistic PCGML approaches represent a generative space by modeling how likely different interactions and combinations are. These probabilities are typically estimated from a set of training examples from the domain intended to be modeled. In this chapter we discussed both unconditional and conditional probabilities, and showed how these different types of probabilities can be used to model a platformer level. We also discussed Markov models and Bayesian networks, both of which leverage conditional probabilities, and have been used in the past for PCGML. In Chap. 8 we will introduce some methods that extend the Markov chain approaches by greatly extending the context of the model, which can lead to increasingly sparse models. We discussed most approaches in the context of platformer level generation, partially for ease of explanation and consistency, but also because the approaches presented have been used most commonly in that domain. In Chap. 12 we will discuss how one of the open problems in PCGML is lack of diversity in testing and application domains. However, for now we will note that the graphical models discussed in this chapter can be applied to domains easily represented as graphs (e.g., buildings, stories, quests, rules, etc.).

In this chapter we also introduced latent variables, which are an important concept when trying to model complex domains and particularly in domains with hidden information. As noted, clustering is one way of identifying such variables; and clustering has been used in a variety of PCGML approaches for trying to identify groups of structures and other high-level patterns. The following few chapters will further explore the use and power of latent variables as they apply to neural networks.

Neural Networks—Introduction

In Chap. 3 we learned about linear regression, a model where we learn a set of weights to apply to our input, in hopes of fitting a linear model that produces the lowest Mean Square Error.

$$\hat{Y} = Wx + b \tag{7.1}$$

This simple model will be our window into neural networks, a class of more complex models, but all of the more advanced models come down to the same basic act, finding weights that minimize the error of the model in predicting the training data.

First, let's think of an example where this model could be applicable to PCG. Let's try to emulate the Markov Chain model introduced in Chap. 6. Recall that the model looked at the neighbors of a tile to try to get the probability that the given tile should be of a certain class:

$$P(T_{i,j} = Class | T_{i-1,j}, T_{i,j-1}) \tag{7.2}$$

See Fig. 7.1 for reference. If we wanted to represent this as a regression, we first need to take some steps to convert this categorical representation into numbers, as a regression will multiply our numerical representation with real valued numbers to produce a numerical output. Our first instinct might be to do something like creating a key: $\langle\, 0 : \text{Sky}, 1 : \text{Grass}, 2 : \text{Brick}, 3 : \text{Enemy} : 4 : \text{Powerup}\,\rangle$ Which we could then use as a vector, as shown in Fig. 7.2.

However, this has a problem, in that we are now assuming a mathematical relationship between these discrete classes. Do we really want to say that $Enemy = Powerup - Grass$? Or $Grass \cdot Powerup = Powerup$? There might be mathematical relationships that we would care to learn, but it is naive to assume that basic arithmetic will be useful on arbitrarily chosen numbers for our classes. Instead, we want a representation that holds these discrete classes to be distinct from each other. I.e., $Enemy + Grass \neq Powerup \neq Brick \cdot Brick$.

© The Author(s), under exclusive license to Springer Nature Switzerland AG 2022
M. Guzdial et al.. *Procedural Content Generation via Machine Learning.*
Synthesis Lectures on Games and Computational Intelligence.
https://doi.org/10.1007/978-3-031-16719-5_7

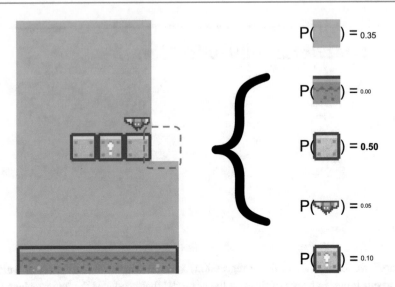

Fig. 7.1 An example of a multi-dimensional Markov chain (as in Chap. 6). The red dotted bubble is the tile that needs to be chosen next, and the probabilities for different classes are shown on the right

Fig. 7.2 Using tile type indices in a vector. The dimensions of the vector correspond to the tiles found to the left and bottom and the values are the indices for those tiles

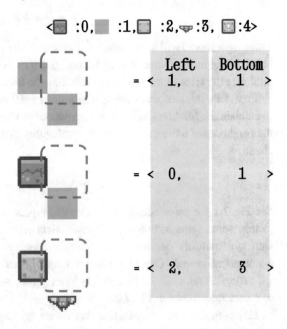

The standard approach for this is what is called a **One-Hot Encoding**. In a One-Hot Encoding each discrete class get its own entry in a vector. In our specific case, we will concatenate two One-Hot Encodings together (one for the horizontal neighbor and one for the vertical neighbor):

And to represent a given configuration, the corresponding entries are 1's, while the rest are 0's (hence the One-Hot name), as can be seen in Fig. 7.3. This now lets us assign a weight to every class entry to predict the class of the next tile. The One-Hot encoding we just discussed will work as a label representation, as well, but in a linear regression we are only producing a single number at the end—if we want to predict from multiple discrete classes, we will need some way to produce multiple numbers (since we do not want to use the naive numerical encoding we dismissed earlier). We will put that aside for the moment, and pretend that we are predicting a single class in our regression. For instance, let's say we want to predict if a given tile is going to be a grass tile.

Just as with the Markov chain, we would step through our level, but instead of counting the number of times we see an occurrence, we are constructing pairs of input, X, and label, Y. Figure 7.4 shows a visual representation of this process. We are now trying to find weights that best minimize:

$$\Sigma||Y - (Wx + b)||^2 \tag{7.3}$$

Fig. 7.3 Representation for the MdMC using the concatenated One-Hot Encodings (denoted as Left and Bottom). Both share the same representation (e.g., Sky is <0, 1, 0, 0, 0> for both)

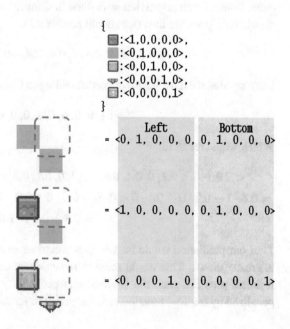

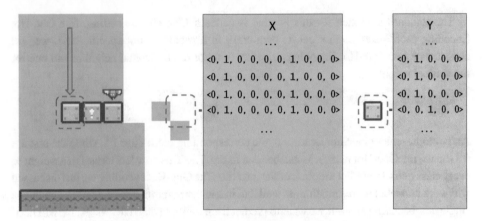

Fig. 7.4 A visual representation of how a level map might be converted into a One-Hot Encoding. Each location in the map is converted (shown mid process). The input (X) consists of the two One-Hot Encodings (Left and Bottom), while the output (Y) consists of the single One-Hot Encoding

This is our **loss** function—the Mean Square Error that we introduced in Chap. 3. Recall, that the loss function tells the model how wrong it is, and the higher the value the more wrong it is. In some ways, we can think of this as the distance between our prediction and the true value (although this falls under the mathematical definition of distance which is only loosely related with our intuitive understanding of distance). Let's pretend that we have the following weights for the prediction of whether the next tile is a grass tile (perhaps they came from a linear regression as defined in Chap. 3) and that the ground truth is that there should be a grass tile next (we should predict a 1):

$$< 0.8, -0.7, 0.1, 0, 0.4, 0.8, -0.2, 0, 0.1, 0.3 > \tag{7.4}$$

Let's say that we have the input corresponding to Grass To the Left and Grass Below:

$$< 1, 0, 0, 0, 0, 1, 0, 0, 0, 0 > \tag{7.5}$$

Which would result in:

$$< 0.8, -0.7, 0.1, 0, 0.4, 0.8, -0.2, 0, 0.1, 0.3 > \cdot < 1, 0, 0, 0, 0, 1, 0, 0, 0, 0 >$$
$$= 0.8 \cdot 1 - 0.7 \cdot 0 + 0.1 \cdot 0 + 0 \cdot 0 + 0.4 \cdot 0 + 0.8 \cdot 1 - 0.2 \cdot 0 + 0 \cdot 0 + 0.1 \cdot 0 + 0.3 \cdot 0$$
$$= 0.8 + 0.8 = 1.6 \tag{7.6}$$

Thus our prediction would be 1.6. This would seem good—we want to predict 1, and 1.6 is greater than 1. This would seem to indicate that our model is extremely confident that it should predict a grass tile, it predicted a probability greater than 100%! However, this is penalized by our loss function as it has gone too far and is starting to deviate from 1:

$$||1 - 1.6||^2 = 0.6^2 = 0.36 \tag{7.7}$$

In fact, this would be the same loss as if the model had predicted a 40% probability that it should be grass:

$$||1 - 0.4||^2 = 0.6^2 = 0.36 \tag{7.8}$$

This is obviously undesirable. We want to predict a high probability that it should be grass, but instead the model is penalized the more confident it becomes. This is due to how the Euclidean loss works—it penalizes as the prediction diverges from the true value, no matter the direction. In this specific example, the Euclidean loss would be minimized only if the weights for Grass Below and Grass to the Left summed to 1 (e.g., both were 0.5):

$$< \mathbf{0.5}, -0.7, 0.1, 0, 0.4, \mathbf{0.5}, -0.2, 0, 0.1, 0.3 > \tag{7.9}$$

Because this weight vector would result in a prediction of 1. But how can we use this information? Intuitively, we know that by changing the weights we can lower the loss. In Chap. 3 we discussed the closed form solution for a linear regression, but that is only one way of finding these weights (and it is only suitable for linear regressions). In this chapter, we will utilize a different, more general approach: **Stochastic Gradient Descent** (SGD).

7.1   Stochastic Gradient Descent

What is SGD? Let's break it down word by word:

- **Stochastic**—The process is done on individual randomly selected (where the term stochasticity comes from) instances of our dataset (as opposed to the entirety of the dataset at once, which would be *Batch* Gradient Descent)
- **Gradient**—A vector field ∇ composed of the partial derivatives of a function with respect to all variables
- **Descent**—We will be following the gradient down (i.e., we will be minimizing our loss) as opposed to Gradient *Ascent*.

At each step of Stochastic Gradient Descent we update our weights in the direction that minimizes the loss for that instance, x_i, from our dataset.

$$w := w - \nabla L(w, x_i) \tag{7.10}$$

Let's return to our previous example, where given an input corresponding to Grass to the Left and Grass to the Right the model predicts 1.6 when instead it should predict 1 resulting in a loss of 0.36. To update our weights with stochastic gradient descent we have to take

Table 7.1 Partial derivatives for the functions we use in calculating the Mean Square Error

Function	Partial derivative
$f(a) = a^2$	$\frac{\partial f}{\partial a} = 2a$
$f(a, b) = a + b$	$\frac{\partial f}{\partial a} = 1$
$f(a, b) = a \cdot b$	$\frac{\partial f}{\partial a} = b$

the partial derivatives of our loss with respect to our weight vector. We use a few functions here, and so review their partial derivatives in Table 7.1:

Starting backwards from our loss, we have:

$$L = (\hat{y} - y)^2 \tag{7.11}$$

$$\frac{\partial L}{\partial \hat{y}} = 2(\hat{y} - y) \tag{7.12}$$

$$\hat{y} = \sum_i w_i \cdot x_i \tag{7.13}$$

$$\frac{\partial \hat{y}}{\partial w_i} = x_i \tag{7.14}$$

To put it all together, we use the Chain Rule of Calculus which states:

$$\frac{\partial z}{\partial x} = \frac{\partial z}{\partial y} \frac{\partial y}{\partial x} \tag{7.15}$$

Doing so, we have:

$$\frac{\partial L}{\partial w_i} = \frac{\partial L}{\partial \hat{y}} \frac{\partial \hat{y}}{\partial w_i} = 2(\hat{y} - y)x_i \tag{7.16}$$

This gives us the gradient of our loss with respect to our weights. In our case, this would be:

$$\frac{\partial L}{\partial w_0} = 2(1.6 - 1)1 = 1.2 \tag{7.17}$$

This means that at the current point, increasing w_0 by 1 would *increase* our loss by 1.2. Since we are doing gradient *descent* we want to go in the opposite direction, so we want to negate our gradient and then step in that direction:

$$w_i := w_i - \nabla L(w_i, x_i) \tag{7.18}$$

$$w_i := w_i - 2(\hat{y} - y)x_i \tag{7.19}$$

In our case, this would be:

$$w_i := w_i - \nabla L(w_i, x_i) \qquad (7.20)$$
$$w_0 := 0.8 - 1.2 = -0.4 \qquad (7.21)$$

Given that only two of our inputs our non-zero (indices 0 and 5), and that the weights for both entries are 0.8, our weights afterwards would be identical except entries 0 and 5 would now be -0.4. If we were to try that training instance again we would get a loss of:

$$||1 - -0.8||^2 = 1.8^2 = 3.24 \qquad (7.22)$$

Uh oh—not only did our loss not get better, it got a lot worse. This is because we shouldn't directly follow the gradient and instead should use a **learning rate**—η (sometimes referred to as a **stepsize** and also sometimes represented as α) that will attenuate our gradient descent steps. Generally, a gradient gives us a direction (and magnitude) to update our weights towards, not a value to change our weights to exactly. The magnitude corresponds to the relative degree that we want to change the weights, but is not an exact value corresponding to how to change the weights. Instead of:

$$w_i := w_i - \nabla L(w_i, x_i) \qquad (7.23)$$

We would use:

$$w_i := w_i - \eta \nabla L(w_i, x_i) \qquad (7.24)$$

This η will typically be small, say 0.001 or so. In this simple example we will use a larger stepsize of 0.1 to see a larger impact:

$$w_i := w_i - \eta \nabla L(w_i, x_i) \qquad (7.25)$$
$$w_0 := 0.8 - 0.1 \cdot 1.2 = 0.68 \qquad (7.26)$$
$$L = ||1 - 0.68 \cdot 2||^2 = 0.36^2 = 0.1296 \qquad (7.27)$$

So our loss has gone down from 0.36 to 0.1296. If we were to do this again, we would get:

$$\Delta w_i = 2(1 - 0.68 \cdot 2) = 0.64 \qquad (7.28)$$
$$w_0 = 0.68 - 0.1 \cdot 0.64 = 0.616 \qquad (7.29)$$
$$L = ||1 - 0.616 \cdot 2||^2 = 0.232^2 = 0.053824 \qquad (7.30)$$

which again is a decrease! We could keep doing this until our loss is 0 (or does not improve). We see that indeed, our weights are getting closer and closer to 0.5 (which would result in a prediction of 1). However, given that a tile is highly likely to be a grass tile if both the tile to the left and below it are also grass, it doesn't seem right to penalize being overly sure that the tile should be grass. Unfortunately, this is a consequence of setting up our problem with a linear regression—we are trying to get close to a real valued number (either a 0 or 1 in this case) and we are penalized if we go too far. However, we originally set out to emulate a

Markov chain, except as a regression. The Markov chain provides probabilities for the class of the next tile, so we want to set up our regression in a way that emulates this. Furthermore, we would like it if our model wasn't be penalized for being too certain about its predictions. If it predicts a higher value for a positive outcome (or conversely, a lower value for a negative outcome), it should have lower loss, not higher.

7.2   Activation Functions

Instead of predicting a real valued number that can take any value between $(-\infty, \infty)$, we need an **activation function** that will constrain our output to be suitable as a probability. To recall, a discrete probability function must have all values between 0 and 1 (inclusive) and the sum over all possible events must be 1. While there are infinitely many functions that can satisfy those constraints, the most commonly used one is the **logistic sigmoid** function (as visualized in Fig. 7.5)

$$\sigma(x) = \frac{1}{1 + e^{-x}} \tag{7.31}$$

The logistic sigmoid is a function whose domain is $\mathbb{R} \to (0, 1)$. It is also symmetric, so $1 - \sigma(x) = \sigma(-x)$, which means that the probabilities for an event occurring and it not occurring sum to 1. Furthermore, while it is beyond the scope of this book, it is also intimately linked to the odds ratio of an event occurring (i.e., $\frac{p}{1-p}$, see [14]).

However, we have an issue—Mean Squared Error does not make sense if we are trying to predict a probability value. We will still want to maximize the likelihood, but we need a different loss function given our assumptions. If the estimated probability of output i is q_i, which is seen p_i times in the training set (of size N) then the likelihood of the model given the training set is

Fig. 7.5 The Logistic Sigmoid function. It is so named because of its S shape (sigma is Greek for S)

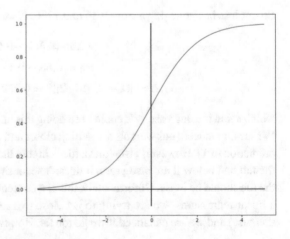

$$\prod_i q_i^{p_i} \tag{7.32}$$

which if we take the logarithm of we get

$$\sum_i p_i \log q_i \tag{7.33}$$

which is the negative Cross-Entropy between the distributions p and q:

$$CrossEntropy(p, q) = -\sum_i p_i \log q_i \tag{7.34}$$

To maximize the likelihood of our model given the data, we must minimize the Cross-Entropy, so Cross-Entropy is the loss for our Logistic Regression. So, with Stochastic Gradient Descent and Logistic Regression we can finally emulate the Markov chain introduced earlier, right? Well, remember, the Logistic Regression only emulates a **Bernoulli distribution** (a distribution for a single yes/no coin toss), but we want a **Categorical distribution** (a distribution for a single choice between many different outcomes like picking a single card out of a deck). To do so, we need to change our framing. We could do multiple logistic regressions—one for each class—but this still wouldn't let us randomly pick from between the different outcomes. E.g., if we randomly sample from each class, what do we do if it says that a tile should simultaneously be grass and sky? Instead, we will will return to our earlier framing:

1. The values must fall within the range of 0–1
2. All values must sum to 1

But instead of only having two values, we need a probability value for multiple outcomes. To do so, we will use the Softmax function—a function that produces probabilities for a multi-class distribution:

$$Softmax(x_i, X) = \frac{e^{x_i}}{\sum_{x \in X} e^x} \tag{7.35}$$

We see that if $x \in X$ are all real valued then e^{x_i} is in the range $[0, \inf]$. Since we are dividing by the sum of all possible outcomes, then this constrains the values in the Softmax to be in [0–1] meeting (1). We can also see that:

$$\sum_{x \in X} Softmax(x_i, X) = \frac{\sum_{x \in X} e^x}{\sum_{x \in X} e^x} = 1 \tag{7.36}$$

So (2) is met. The only issue is that we are changing from a vector of weights $W_{1,k}$ to a matrix of weights $W_{c,k}$ where c is the number of classes and k is the **dimensionality** of our input. In our current example we have 5 classes (Sky, Grass, Brick, Enemy, Powerup) so $c = 5$ and given that we are looking at two tiles for input $k = 2 \cdot 5 = 10$. Dimensionality is just another way to refer to the number of entries in our vector—thinking in two dimensional space we have two entries in a vector (horizontal and vertical) and three dimensional space has three (horizontal, vertical, and depth). Thus our regression now produces c different values, called **Logits**, which we convert into probabilities by applying the Softmax function. We will need an appropriate loss function, and we turn again to the Cross Entropy loss.

Is it possible that we can finally emulate the Markov chain introduced earlier with Stochastic Gradient Descent and Softmax Regression? For certain configurations we can, but con-

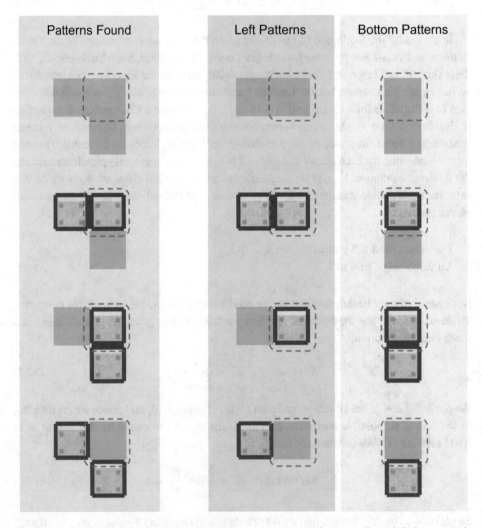

Fig. 7.6 An example of the Exclusive-Or problem as represented in a case that might be present in an MdMc

sider the following example shown in Fig. 7.6. There is no possible combination of weights for our Softmax regression that would be able to make these predictions correctly. If we assign a high weight for Sky on the left → Sky then, the 3rd instance does poorly, but conversely if we assign a low weight, then the first does poorly. Similarly, Sky from below has issues with the 1st and 2nd. So too for Bricks on the left for the 2nd and 4th and Bricks below for the 3rd and 4th. This is an example of the **Exclusive-Or problem**—a toy problem that demonstrates the issue with Linear models—models whose outcomes are determined by a linear transformation of the input. Without considering the **interaction** between the two features—e.g., It should be sky when both left and bottom are the same and bricks when they are different—then it is impossible for the model to learn the pattern. A linear model treats the features independently, without any chance for patterns involving multiple features to be learned. To deal with such cases (and more complicated) we will need a model that is capable of non-linear behavior, and for that we turn to **Artificial Neural Networks**.

7.3   Artificial Neural Networks

Artificial Neural Networks are a class of model that can learn arbitrary non-linear behavior. Classically referred to as ANNs (because NN referred to Nearest Neighbors) but more recently the artificial is dropped and people refer to them as NNs. In fact, the Universal Approximation Theorem [32, 78] states that a NN with a single layer of arbitrary width can approximate any continuous function to arbitrary precision. Obviously, this seems—and is—very powerful. If we have a function that we wish to approximate—e.g., learn the function that: produces levels similar to a set of example levels, generates new *Magic: The Gathering* cards, produces new rules for variants of Poker, etc.—then, *theoretically* a NN is capable of doing so. Note, just because it is theoretically possible does not mean that it is practical—or even possible—to achieve. Nonetheless, *how is this possible*? Let's first break down what exactly is meant by a Neural Network.

Classically, an artificial neural network is composed of layers—see Fig. 7.7. Each layer is composed of a number of "**neurons**" (this is where the name *Neural* Network comes from) that combine their inputs and then produce an "activation"—using the neuron metaphor. These activations get passed on as the input to the next layer, and so on until the final output. See Fig. 7.7 for an example. Demystifying the metaphor, each neuron is actually very similar to the regressions introduced earlier in the chapter. In the most general form, going from individual neurons to entire layers:

$$o_{ij} = a(w_{ij} \cdot O_{i-1}) \tag{7.37}$$

$$O_i = a(W_i O_{i-1}) \tag{7.38}$$

o_{ij} = The output of the jth neuron in the ith layer

O_i = The output of the the ith layer

w_{ij} = The weight vector applied to the $i - 1$th layer for the jth neuron.

W_i = The weight matrix applied to the output of the $i - 1$th layer

O_i, O_{i-1} = The output of the ith and $i - 1$th layers

a = The activation function applied to the intermediate output

Most simply, this would be a neural network with a single hidden layer:

$$h = a(W_h \cdot x) \tag{7.39}$$

$$o = a_o(W_o \cdot h) \tag{7.40}$$

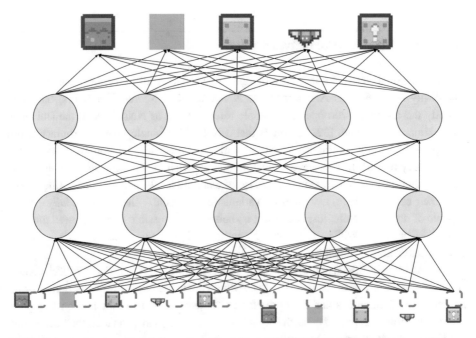

Fig. 7.7 A graphical representation of a Neural Network (NN). This NN continues our Softmax prediction of tiles for a Markov model. The input vector is at the bottom and the output vector is at the top. The edges between *neurons* represent the weight applied to the given input for the respective output. Each inner—*hidden*—neuron would have a non-linear activation that it applies to the sum of its input

As noted, this is very similar to the regressions that we have discussed up to this point—we are taking the dot product of a weight vector and an input. The only distinction is that there is an activation function a being applied. For the NN to be capable of learning arbitrary functions efficiently, a must be a suitable (i.e., non-polynomial, differentiable—the importance of these will be discussed below) activation function (and there might be a different output activation function a_o). Most commonly, each neuron produces a single value ($n_i = 1$), meaning that the weight matrix for each neuron is a vector—just as in our earlier regressions. Looking to Fig. 7.7, each line represents a scalar component of w_{ij} with the entire vector being all of the lines feeding in to a given neuron. Therefore, all of the lines between two layers actually just represent a matrix. However, since each layer of the neural network commonly contains multiple neurons, we commonly think in layers and not individual neurons, so the ith layer is:

$$O_i = a(W_i O_{i-1}) \tag{7.41}$$

$$w_i = \text{The weight matrix applied to the output of the } i - 1\text{th layer}$$

where the only difference from above is that W_i is a $\mathbb{R}_{d_i \cdot d_{i-1}}$ weight matrix, where d_i is the number of neurons found in layer i. a is assumed to also have been changed from $\mathbb{R}^{n_i} \rightarrow \mathbb{R}^{n_i}$ to $\mathbb{R}^{d_i} \rightarrow \mathbb{R}^{d_i}$), but in most cases, the activation function is applied element-wise to the input, so the dimensionality is unimportant. In the case of Fig. 7.7, d_0 (the dimensionality of the input) $= 10$, and $d_1, d_2, d_3 = 5$.

As a note, so far, we have been focused on the hidden neurons and layers. The output layers are similarly composed, but are more akin to the earlier regressions. Instead of a non-linear activation function, we instead have an output activation, see Table 7.2 for more details.

In essence, each hidden layer of a neural network is a regression from an input of a certain size to an intermediate output of a certain size. However, one key difference between a NN and a regression is the presence of the activation function in the hidden layers. The activation function is what provides the potency of a neural network. Imagine for a minute

Table 7.2 Table of some commonly used output activation

Output Function	Range	Predicts	PCGML Uses
Identity	$(-\infty, \infty)$	A $\mathbb{R}$ number	Rarely applicable
Sigmoid	$(0, 1)$	Binary outcome	Something there or not?
Softmax	$(0, 1)^d$	Probability of class $= d$	Which thing is there?

that the activation was the Identity function—a number goes in and comes out the same. Let's assume that we have a neural network with 1 hidden layer (meaning we have two layers, Input → Hidden and Hidden → Output). In this case, our network would look like:

$$Y = W_o \cdot a(W_h \cdot x) \tag{7.42}$$

but since a is the Identity function, this is actually:

$$Y = W_o \cdot W_h \cdot x \tag{7.43}$$

$W_o \cdot W_h$ is just a matrix multiplication and can be factored into a single matrix, $W_o \cdot W_h = W$. This means that our "deep" network with a hidden layer is just a refactored linear regression, and this can be extended to any number of layers. Commonly, modern neural networks have many hidden layers making them "deep" (in contrast to earlier neural networks that had a single hidden layer). But, we see that the "deep" portion of the neural network is unimportant without a non-polynomial activation.

But what makes a good activation function? There are two requirements for an activation function:

1. It is non-polynomial
2. It is differentiable (or at least sub-differentiable)

The first requirement we have already covered, and the second is related to issues we discussed earlier. Recall that previously we discussed the use of Stochastic Gradient Descent by which we update the weights of our model by finding the gradient of the loss with respect to the weights—to train our neural network we use the same process. To do so, we must calculate the loss with respect to the weights, and the only way that we can guarantee that this works is if our activation functions are differentiable. By constructing our neural network out of differentiable components we can use the chain rule to calculate the gradient of the loss with respect to a weight value, no matter where it is in the network, via a process known as **Backpropagation**. The name Backpropagation comes from 1986 [146] but has existed as a concept since the 1960s [91], the idea being that if a function is composed of differentiable sub-functions then the chain rule of calculus can be applied to find the derivative with respect to any of the given sub-functions. To recall, the chain rule of calculus states:

$$h(x) = g(f(x)) \rightarrow \frac{\partial h}{\partial x} = \frac{\partial g}{\partial f}\frac{\partial f}{\partial x} \tag{7.44}$$

By following the chain of partial derivatives we can calculate the gradient of the loss with respect to any parameter in the neural network, allowing us to train the network via SGD. Beyond the requirements of non-polynomial and differentiable, the implications of a given activation function come down to its effect on the output and the gradient. Table 7.3 shows these for a wide variety of activation functions.

Table 7.3 Table of activation functions commonly used in hidden layers of a NN

Activation	Linearity	Gradient	Time Cost
$a + b$	Linear	Passes freely	–
$a \cdot b$	Linear	Is multiplied by the complement	–
$\sigma(a)$	Non-Linear	Flat for extreme values	Expensive
$tanh(a)$	Non-Linear	Flat for extreme values	Expensive
$max(a, b)$	Non-Linear	Only to the maximal element	–
$ReLU(a)$	Non-Linear	Passes freely for values >0	Cheap

Having gone through an introduction to the inner workings of Neural Networks, now let's consider a few introductory PCGML cases using NNs. In the following two chapters we will discuss more advanced models, but for now we will consider a few different input/output modalities for levels and how they can be implemented with the NN basics we have covered in this chapter.

7.4 Case Study: NN 2D Markov Chain

Let's return to the Markov chain example that we have been working through. First, we need to determine what our input and output are going to be. Figure 7.4 shows how we will go from input (neighbor tile classes) to output (class of predicted tile). Algorithm 7.1 describes in pseudocode the entire training and sampling procedure for a Neural Multi-dimensional Markov chain.

In lieu of the One-Hot Encoding discussed earlier, we will instead use an **Embedding**. To motivate this, consider that in a dot product involving a One-Hot encoding of dimensionality n there is one multiplication that matters $(1 \cdot a)$ and $n - 1$ that don't $(0 \cdot a)$ (and none of the additions matter). Note, for our concatenated representation there would be two multiplications with 1s and $2n - 2$ with 0s. An embedding layer removes the dot product and instead we just input the index into a table (the entry that would have been a 1) and get the associated row. The end result of using an embedding is the same as with a One-Hot encoding, but it is more computationally efficient. We will also use a **Vocabulary**—a mapping from a tile to its index. Thus for our input we would take all of the multi-dimensional neighbors of the tile we wish to predict, find their unique indices via the Vocabulary, and then find the embeddings corresponding to those indices.

We will then concatenate these embeddings together to produce the input to the NN (this entire procedure occurs on lines 19 and 29 for Training and Usage, respectively). We will then

---

Algorithm 7.1 Neural MdMC

1: {**Definitions**}
2: V //Vocabulary of Tile Types
3: $v = |V|$ //SizeofVocabulary
4: n //Context neighborhood of MdMC
5: k //Number of Hidden Layers
6: d //Size of Hidden Layers
7: a //Hidden Activation
8: {**Initialization**}
9: **for** $i = 1..n$ **do**
10:     $E[i] \leftarrow$ Embedding(v)
11: **end for**
12: $W_1 \leftarrow$ Linear($v \cdot n, d$) //From |context|*|vocab| to |hidden|
13: **for** $i = 2..k - 1$ **do**
14:     $W_i \leftarrow$ Linear(d, d) //From |hidden| to |hidden|
15: **end for**
16: $W_k \leftarrow$ Linear(d, v) //From |hidden| to |output|
17: {**Training Step**} $- t$ //The current tile to predict
18: $T \leftarrow$ array()
19: $T \leftarrow [E[i](V(c))$ **for** $i, c \in$ enumerate(Context(t))]
20: $x \leftarrow T$
21: **for** $i = 1..k - 1$ **do**
22:     $x \leftarrow a(W_i \cdot x)$
23: **end for**
24: $p \leftarrow$ SoftMax($W_k \cdot x$) //Predicted probabilities
25: $l \leftarrow L_{CE}(p, t)$//Loss is Cross Entropy between predictions and truth
26: backward(l)
27: step(∇l)
28: {**Usage Step**} $- t$ //The current tile to predict
29: $T \leftarrow [E[i](V(c))$ **for** $i, c \in$ enumerate(Context(t))]
30: $x \leftarrow T$
31: **for** $i = 1..k - 1$ **do**
32:     $x \leftarrow a(W_i \cdot x)$
33: **end for**
34: $p \leftarrow$ SoftMax($W_k \cdot x$) //Predicted probabilities
35: **return** Sample(p)

---

have some number of other hidden layers (Initialized on lines 13–15, Used on lines 21–23, 31–33). Finally, we need to use a Softmax (lines 24 and 34) to turn our hidden representation into a categorical probability distribution. When training, we then calculate the Categorical Cross Entropy Loss between the predicted probabilities and the truth (line 25), calculate the gradient of the loss (line 26), and then change (commonly referred to as **stepping** the optimizer) our weights via the gradient of the loss (line 27). In the course of training we will do this many, many times. The two main ways of doing this art to either run for a set number of steps or for a set number of passes over the data (referred to as an **epoch**). When using the trained model for generation, we instead sample from the probability distribution (line 35). After training for some number of epochs and the loss has converged, we can use the learned NN Markov Model in lieu of the tabular one found in Chap. 6. It might seem that the input representation would limit the model. For the tabular MdMC, <Grass left, Grass below> has no relation to <Empty left, Grass below>, while the NN would use the same weights for "Grass below." However, so long as the weight matrices are of sufficient size, the Universal Approximation Theorem states that the NN *can* learn the unique combinations, leading to the exact same performance. The only difference between the training procedure and the use of the learned NN is that instead of calculating the loss and performing backpropagation (lines 28, 29, and 30), the predicted probabilities are sampled (line 42).

7.5   Case Study: NN 1D Regression Markov Chain

Of course, there are plenty of other ways we could formulate our neural Markov model. One approach from Hoover et al. does not predict tile by tile, but instead reframes the problem as a regression problem where the highest height for a tile type (in their case solid ground) is predicted for each column [77]. This is still Markovian in that it is based only the current state (which may be based on multiple neighbors). First, let's determine our input and output exactly.

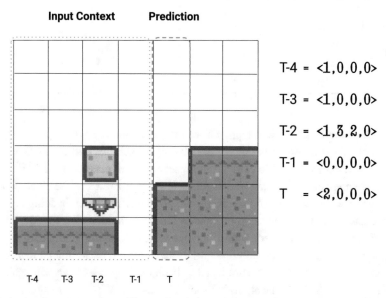

Fig. 7.8 A visual representation of how a level map might be converted into a height map. Each column in the map is converted into a vector where each tile type has its maximum height recorded

---

Algorithm 7.2 Neural Regression

1: {**Definitions**}
2: V //Vocabulary of Tile Types
3: $v = |V|$ //Size of Vocabulary
4: n //Number of columns in context
5: k //Number of Hidden Layers
6: d //Size of Hidden Layers
7: a //Hidden Activation
8: {**Initialization**}
9: $W_1 \leftarrow \text{Linear}(v \cdot n, d)$ //From |context|*|vocab| to |hidden|
10: **for** $i = 2..k - 1$ **do**
11:    $W_i \leftarrow \text{Linear}(d, d)$ //From |hidden| to |hidden|
12: **end for**
13: $W_k \leftarrow \text{Linear}(d, 1)$ //From |hidden| to 1
14: {**Training Step**} $- h$ //The height to predict
15: $H \leftarrow [\text{height(i,c)} \textbf{ for } i, c \in \text{Context}(t)]$
16: $x \leftarrow H$
17: **for** $i = 1..k$ **do**
18:    $x \leftarrow a(W_i \cdot x)$
19: **end for**
20: $l \leftarrow L_{MSE}(x, h)$ //Loss is Mean Square Error between predictions and truth
21: backward(l)
22: step(∇l)
23: {**Usage Step**} $- h$ //The height to predict
24: $H \leftarrow [\text{height(i,c)} \textbf{ for } i, c \in \text{Context}(t)]$
25: $x \leftarrow H$
26: **for** $i = 1..k - 1$ **do**
27:    $x \leftarrow a(W_i \cdot x)$
28: **end for**
29: return x

Somewhat akin to the earlier approach we need each index in our input to correspond to a given class and timestep, but unlike the other approach this will not be One-Hot. Just as before the value will be 0 if a class is not present (e.g., all classes at T-1 in Fig. 7.8), but unlike before non-zero values can be something other than 1 (e.g., at T-2 block = 3, enemy = 2, and grass = 1). Because of this change we no longer use an embedding layer for the different classes, but the same setup of input to hidden to hidden to ... to hidden to output will remain the same. The other major difference is that we are not doing classification, so we do not need an output activation function. Instead, we are performing a regression—in fact the same regression that we introduced to back in Chap. 3—so we will again use the Mean Square Error (MSE). The algorithm shown in Algorithm 7.2 is very similar to the one in Algorithm 7.1, with a few small distinctions.

1. There is no embedding, instead just using the heights which are already scalar values
2. The predicted value is the height, a single scalar number, so there is no Softmax activation on the output.

7.6   Case Study: NN 2D AutoEncoder

One thing you will notice is that both of the previous examples generated the content a little bit at a time. These were **Autoregressive** models (autoregressive meaning that the current output of the model is based on its previous output). This is a common structure for many forms of PCGML, but it is not the only way. Another common form is to go from some **latent** representation (a hidden representation that we believe is *latent* within the artifacts we wish to learn from) to the output that we want in one shot. To do so, we need to both learn a way of going from the input to this latent representation and a way to go from the latent representation to the output. We will do this by constructing an **AutoEncoder**—a type of model that has an **Encoder** (data $\to$ latent) and **Decoder** (latent $\to$ data). One of the major aspects of an AutoEncoder is that the size of the latent representation is typically much smaller than the size of the original data – this dimensionality reduction is important for learning a dense latent space with meaningful features (while it is possible to have a latent space the same size or larger than the data, this is typically unadvised). Figure 7.9 shows a visual representation of an AutoEncoder.

Like before we will use a tile based representation, but instead of predicting one tile at a time, we will predict all tiles simultaneously. Meaning we will predict the tiles at $< 0, 0 >$, $< 1, 0 >, ... < w, h >$ all at once. Metaphorically, we might think of this as an image, where we are predicting the *color* (tile) for each *pixel* (location). As before we will use an embedding for the tiles as the first layer, but we will share the embedding across all individual locations in the level that we wish to generate (in contrast, the previous two examples used

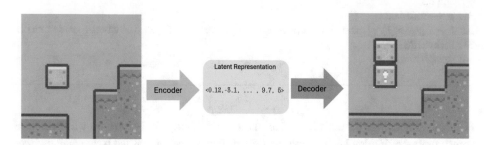

Fig. 7.9 A visual representation of how an AutoEncoder might be used to generate a level. The AutoEncoder has an Encoder (blue) that maps from input to the latent representation (yellow), and a Decoder (red) that maps from latent representation to output. Note: this is a form of *lossy* compression, so the generated artifact corresponding to a given latent representation might not perfectly recreate the original artifact, as seen in this example

---

Algorithm 7.3 AutoEncoder Initialization and Training

1: {**Definitions**}
2: V //Vocabulary of Tile Types
3: $v = |V|$ //Size of Vocabulary
4: n //Number of entries in input (width x height)
5: k_E //Number of Hidden Layers for Encoder
6: k_D //Number of Hidden Layers for Decoder
7: d_E //Size of Hidden Layers for Encoder
8: d_L //Size of the Latent Representation
9: d_D //Size of Hidden Layers for Decoder
10: a //Hidden Activation
11: {**Initialization**}
12: $E \leftarrow \text{Embedding}(v)$
13: $W_1^E \leftarrow \text{Linear}(v \cdot n, d_E)$ //From |context|*|vocab| to |hidden| for Encoder
14: **for** $i = 2..k_E - 1$ **do**
15:     $W_i^E \leftarrow \text{Linear}(d_E, d_E)$ //From |hidden| to |hidden| for Encoder
16: **end for**
17: $W_{k_E}^E \leftarrow \text{Linear}(d_E, d_L)$ //From |hidden| to |latent|
18: $W_1^D \leftarrow \text{Linear}(d_L, d_D)$ //From |context|*|vocab| to |hidden| for Decoder
19: **for** $i = 2..k_D - 1$ **do**
20:     $W_i^D \leftarrow \text{Linear}(d_D, d_D)$ //From |hidden| to |hidden| for Decoder
21: **end for**
22: $W_{k_E}^E \leftarrow \text{Linear}(d_E, v \cdot n)$ //From |hidden| to |context|*|vocab|
23: {**Training Step**} – I //The "image" to predict
24: $x \leftarrow [E[i]i \in I]$
25: **for** $i = 1..k_E$ **do**
26:     $x \leftarrow a(W_i^E \cdot x)$
27: **end for**
28: **for** $i = 1..k_D$ **do**
29:     $x \leftarrow a(W_i^D \cdot x)$
30: **end for**
31: $p \leftarrow \text{SoftMax}(x)$
32: $l \leftarrow L_{CE}(p, I)$ //Loss is Cross Entropy between predictions and truth
33: backward(l)
34: step(∇l)

---

different embeddings for each individual facet of the input). As before, we will then have some number of hidden layers, with the only difference being that we have some number of hidden layers in the Encoder and some in the Decoder, with the middle being our latent representation. Algorithm 7.3 shows the training setup for an AutoEncoder, and Algorithm 7.4 shows the three uses of an AutoEncoder (Encoding, Decoding, AutoEncoding).

Algorithm 7.4 AutoEncoder Usage

1: {**Definitions**}
2: k_E //Number of Hidden Layers for Encoder
3: k_D //Number of Hidden Layers for Decoder
4: a //Hidden Activation
5: {**Encoder Usage**} $- I$ //The "image" to encode
6: $x \leftarrow [E[i] | i \in I]$
7: **for** $i = 1..k_E$ **do**
8:    $x \leftarrow a(W_i^E \cdot x)$
9: **end for**
10: return x
11: {**Decoder Usage**} $- L$ //The latent encoding to decode
12: $x \leftarrow L$
13: **for** $i = 1..k_D$ **do**
14:    $x \leftarrow a(W_i^D \cdot x)$
15: **end for**
16: $p \leftarrow \text{SoftMax}(x)$
17: return $\text{sample}(p)$
18: {**AutoEncoder Usage**} $- I$ //The "image" to encode
19: $x \leftarrow [E[i] | i \in I]$
20: **for** $i = 1..k_E$ **do**
21:    $x \leftarrow a(W_i^E \cdot x)$
22: **end for**
23: **for** $i = 1..k_D$ **do**
24:    $x \leftarrow a(W_i^D \cdot x)$
25: **end for**
26: $p \leftarrow \text{SoftMax}(x)$
27: return $\text{sample}(p)$

AutoEncoders have some desirable properties for PCGML. As a whole, the AutoEncoder can take in content and attempt to reproduce it; however, given the lossy aspect, it can often be beneficial if this reproduction is different. In the work of [83] they used this property of AutoEncoders to "repair" broken content, since the generated content would more commonly not be broken. Utilizing the Encoder, we can encode a piece of content and then sample near it to generate similar, but not identical content. It can also be used for vector analogies. To provide an example in a different domain, if we are able to encode words as vectors, such that the encodings capture meaning, then we can do something like $king - man + woman = queen$. In our level generation case, we might have two identical levels except one is full of enemies and the other has none. We could then find the vector for "enemies" by calculating $enemies = full - empty$. We can then use this vector to modify other levels by adding it to their encodings and then decoding the analogized vector. Finally, the latent representation can be sampled (or explored programmatically) and fed into the Decoder to generate new content.

7.7   Takeaways

In this chapter we went over the theoretical and practical underpinnings of **Artificial Neural Networks**. We discussed the ways in which they are similar to the models that we have discussed previously in Chaps. 3 and 6—they can predict numerical data or probabilities. We also discussed the ways in which they are different—their ability to capture non-linear behavior. We finished with three example uses of NNs—a Neural Markov chain, a Neural Height Predictor, and a Neural AutoEncoder. In the following chapters we will discuss NNs designed to operate on sequences (Chap. 8) and images (Chap. 9).

Sequence-Based DNN PCGML

8

In the previous chapter, we discussed neural networks, building up from a simple model with no layers (i.e., linear regression), to a model with a single hidden layer (two layers in total, hidden + output), and finally discussing 'deep' neural networks with many hidden layers. We also discussed different factors in choosing the representation of our data. In this chapter, we will be discussing a specific representation—sequences—and types of NNs specifically designed to handle sequences.

We might first ask, "Haven't we already done things in sequence?" and that would be correct. The example laid out in Sect. 7.4 treats the level as a sequence of vertical slices that we are predicting one-at-a-time. But the models that we used are *Markovian* (see Chap. 6)— they have no history. They assume that in a given state, it makes no difference how that state was reached and that the state is all that matters. Put another way, they are a type of function where if we put the same input in we will always get the same output, regardless of the input that came before. As a note, the output we get is a probability distribution that we sample from, so the "output" of our generator might differ, but the probability distributions won't. This seems very reasonable, but it limits the ability to handle sequences effectively. Let's walk through an example to gain an intuition for this.

Let's consider an example of generating cards for a collectible card game, like *Magic: The Gathering* or *Hearthstone*. If we wanted to generate cards like this, we might do as we did in Chap. 6's quest example and treat them as a sequence. Colloquially, we might think of them as a sequence of *words*, but more commonly we will think of them as a sequence of **tokens** (individual, atomic pieces of our sequence). Tokens can be anything from individual bytes, to characters, to words, up to combinations of words. In our case, we will treat tokens as words or unique symbols (like a sword). In this way, we could train a model that predicts the next token to be generated. If our training data consisted only of the two cards in Fig. 8.1b, given the token 'Knight' it would learn to generate 'of' (since both times we see 'Knight' it is followed by 'of') and given 'of' there would be a 50/50 chance of generating 'Procgenmel'

M. Guzdial et al.. *Procedural Content Generation via Machine Learning*.
Synthesis Lectures on Games and Computational Intelligence,
https://doi.org/10.1007/978-3-031-16719-5_8

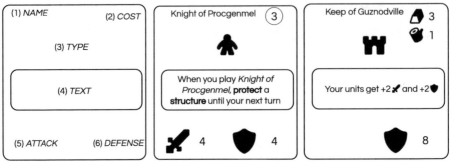

(a) Structure of the cards.   (b) Two cards from a fictional collectible card game.

Fig. 8.1 Examples for card generation task. Assets taken from the Kenney Board Game Icons Pack

or 'Guznodville' (since we have seen 'of Procgenmel' and 'of Guznodville'). For something like the name, that ambiguity might be fine, but for something like the mechanical properties, this ambiguity could be troublesome. For instance, after 'your' there is a 50/50 chance of 'next'/'units'. Considering this, if we were to generate the next token we might wind up generating text like:

- 'you play Knight of Procgenmel, protect a structure until your units get +2'
- 'your next turn'

Maybe the first one is acceptable in the game, but the second is most certainly a nonsensical sentence fragment. One option is that we could just extend our state to include this context—i.e., create a meta state that is a composite of multiple states (in fact this is what is done in Chap. 6). However, this just defers our problem. The size of our context is still limited to a fixed amount, and ideally, we would be able to handle contexts of any arbitrary size. We might want to remember the word 'Knight' when generating text that occurs much further away, and we can't just keep arbitrarily increasing the context size. For one, as we increase the size of the model we will be looking at sparser and sparser patterns.

To put this in real terms, in the complete works of William Shakespeare there are 29,947 unique tokens (words and punctuation). If we look at all combinations of pairs of tokens (known as **bigrams**), there are 330,093 found in the works. However, if we considered all possible pairs $(29,947^2)$ there are 896,822,809 possible. This means that only 1 in 2717 possible bigrams shows up in the actual text. As we increase the size of the context, it gets sparser and sparser. For a context of 4 words the actual count compared to the possible count is 1 out of over 800 billion. This means that while there are a huge number of possible sequences, only a tiny fraction are found in our training data. This might not seem like a problem, but it means that for larger context sizes we are going to be more and more likely to plagiarize from the training set. For instance, in the case of Shakespeare with a context of size 4 the average number of possible tokens that can follow is 1.05, and only 3% of the size 4 sequences have more than one possible token that can follow. This means that if we have a context of size 4 we are going to be pretty much plagiarizing Shakespeare—for the sake of learning the cards we are most likely going to have far less text with much less variety, meaning it would be even worse.

To combat these issues, we need a model that is capable of handling large contexts in a way such that it won't just memorize the training data. To this end, we need to allow our model some way of either (1) remembering the context that it has seen, or (2) allowing it some way to refer back to an arbitrarily sized context. We will now discuss a broad class of models that address (1) by being given a mechanism to 'remember' previously seen input, and (2) we will discuss a bit later in Sect. 8.5.

8.1 Recurrent Neural Networks

In this section, we will discuss **Recurrent** Neural Networks (RNNs). RNNs are neural networks that have a form of memory, allowing them to 'remember' (or at least be influenced by) previous input. The classic single-layered neural network is mathematically formulated as:

$$h = a(W_h \cdot x) \tag{8.1}$$

$$o = a_o(W_o \cdot h) \tag{8.2}$$

$$
\begin{array}{rl}
x = & \text{The input vector} \\
W_h = & \text{The weight matrix for the hidden layer} \\
W_o = & \text{The weight matrix for the output layer} \\
a = & \text{The hidden layer's activation function} \\
a_o = & \text{The output layer's activation function} \\
h = & \text{The output of the hidden layer} \\
o = & \text{The output of the network}
\end{array}
$$

For an RNN, we will be starting from a similar representation. However, the output of our hidden layer will be based on not only the current input but also the output of the hidden layer from the previous step. For the rest of the chapter, there will be references to 'time' steps in relation to RNNs and other models. As a note, these are not necessarily steps in time, but instead are invocations of our RNN sequentially; however, thinking about these steps as occurring in time can be a useful metaphor. Deriving from a recurrent relationship, our RNN has two cases: the base case (the first application of the RNN) and the recurrent case. We will now index our input x, hidden output h, and final output o with the time step (x_i, h_i, and o_i respectively). When defining the base case, it is very similar to the formulation above:

$$h_0 = a(W_h \cdot x_0) \tag{8.3}$$

$$o_0 = a_o(W_o \cdot h_0) \tag{8.4}$$

with the only difference being in the indexing. However, our recurrent case differs:

$$h_t = a(W_h \cdot x_t + W_r \cdot h_{t-1}) \tag{8.5}$$

$$o_t = a_o(W_o \cdot h_t) \tag{8.6}$$

$W_r = $ The weights applied to the hidden output from the previous timestep
$h_t = $ The output of the hidden layer at timestep t
$o_t = $ The output of the network at timestep t

A visualization of an RNN can be seen in Fig. 8.2. In this visualization the RNN has been **unfolded** through time. What is meant by unfolded through time? We can see that the structure of the RNN no longer involves just going from input to output, but instead passes from a node to itself in the future—or thinking in the reverse, the output at a given timestep is dependent on not only the input at that time, but also all previous timesteps. Thinking about it in a different way, we have created a *deep* network, but instead of the depth being an a priori defined aspect of the network, the depth is related to the time. We can see that the "depth" of the network in this case is four, or in other words, it takes four applications of the hidden layer to get from the initial input at t = 0 to the output at t = 3. If we were to construct a non-recurrent network with this same topology, it would be functionally identical

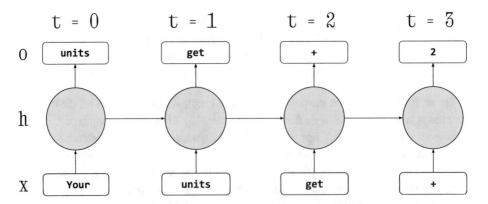

Fig. 8.2 A visual representation of a **Recurrent Neural Network** (RNN). The major difference between the RNN and a standard ANN is the recurrent weight (the edges between the hidden nodes) that allows the RNN to have a state that carries across multiple timesteps

(although one slight oddity is that the weights would be shared between all of the hidden nodes, as that is how the recurrent network works).

Thinking back to the previous chapter, the way that we train a neural network is by following the gradient of the loss backwards through the network and updating the network via backpropagation. In the case of a recurrent network, this gradient is flowing backwards through the network through time—hence, why it is referred to as **backpropagation through time**. In the aforementioned example, the loss at time 3 flows backwards through to the inputs at times 0, 1, 2, and 3. The gradient of the loss as it is applied to the weights for the simple non-recurrent network is:

$$\frac{\partial o}{\partial h} \frac{\partial h}{\partial W_h} \tag{8.7}$$

whereas, the corresponding gradient for the loss at time 3, with respect to the input at time 1 is:

$$\frac{\partial o}{\partial h_3} \frac{\partial h_3}{\partial h_2} \frac{\partial h_2}{\partial h_1} \frac{\partial h_1}{\partial W_h} \tag{8.8}$$

We see that there are two additional steps between the output and the weights. If the partial derivatives are *small* (absolute values between 0 and 1) then the gradient can **vanish** as it flows further backwards in time. If they are *large* (absolute values greater than 1) then they will **explode**. In either case, this is a problem as our gradient will either vanish to nothing, or explode and wreak havoc on numerical stability. This is why "vanilla" RNN's are mostly suitable as a pedagogical tool, and should not be used in practice. Toward this end, two different models have been developed to combat these problems: the **Gated Recurrent Unit (GRU)** RNN and the **Long Short-Term Memory (LSTM)** RNN.

8.2   Gated Recurrent Unit and Long Short-Term Memory RNNs

One of the major issues for both vanishing and exploding gradients is that the gradient is passed backwards through time multiplicatively—which inevitably leads to it growing or decreasing at an exponential rate. The Gated Recurrent Unit Recurrent Neural Network (GRU RNN—or just GRU for short) is a type of RNN developed by Kyunghyun Cho et al. [25] with the intention of solving the vanishing gradient problem. The GRU does so by introducing a few new mechanisms: update and reset gates. These gates are very similar to the hidden states of the RNN, but are constrained to be in the range of 0 to 1. They are then used to affect the flow of the previous step's hidden state to the current step—this means that a portion of the gradient can flow freely backwards through time which helps prevent the vanishing gradient problem. If a standard RNN is of the form:

$$h_t = a_h(W_h \cdot x_t + W_r \cdot h_{t-1}) \tag{8.9}$$

$$o_t = a_o(W_o \cdot h_t) \tag{8.10}$$

the GRU adds some additional computation steps:

$$z_t = a_g(W_z \cdot x_t + U_z \cdot h_{t-1}) \tag{8.11}$$

$$r_t = a_g(W_r \cdot x_t + U_r \cdot h_{t-1}) \tag{8.12}$$

$$\hat{h}_t = a_h(W_h \cdot x_t + U_h \cdot (r_t \circ h_{t-1})) \tag{8.13}$$

$$h_t = (1 - z_t) \circ h_{t-1} + z_t \circ \hat{h}_t \tag{8.14}$$

$$
\begin{aligned}
z_t &= \text{The update gate—used to update the hidden state} \\
r_t &= \text{The reset gate—used to affect the update to the hidden state} \\
a_g &= \text{The gate activation—commonly the logistic sigmoid} \\
a_h &= \text{The hidden activation—commonly } tanh \\
W_z, W_r, W_h &= \text{The weight matrices applied to the input} \\
U_z, U_r, U_h &= \text{The weight matrices applied to the previous hidden state}
\end{aligned}
$$

Note, $\circ$ represents the **Hadamard Product** (i.e., element-wise multiplication). We see that both the update and reset gates of the GRU appear to be very similar to the hidden state of the RNN. The differences come in how the hidden state for the GRU is calculated. We see that the GRU's hidden state is composed of both a candidate hidden state ($\hat{h}_t$) and the previous hidden state (h_{t-1}) where each is multiplied by the value of the update gate (z_t) and its complement ($1 - z_t$), respectively. This means that when the update gate's value is large (close to 1) that the value of the hidden state comes from the candidate hidden state, and, conversely, when it is small (close to 0) that the hidden state is just the previous hidden state.

How does this work to combat the vanishing gradient problem? By using a gate on the hidden state, when the gradient needs to flow backwards through time, it can, because the gate's value can be very, very close to 1. Otherwise, when the gradient doesn't need to flow backwards, it can be close to 0 and the hidden state can be mostly derived from the current input.

8.2.1 Long Short-Term Memory RNNs

The Long Short-Term Memory RNN (LSTM) was developed by Hochreiter and Schmidhuber [74] to tackle the vanishing gradient problem—much like the GRU—except it was developed almost 20 years prior. However, it was not a popular tool for most of those nearly two decades, due to the general unpopularity of Artificial Neural Networks. However, with the advent of GPU aided computation, LSTMs became popular starting around 2015 [87]. In the original work by Hochreither and Schmidhuber, their LSTMs were of a single layer and were limited to just a few LSTM cells (3-4), but by 2015 the LSTMs used by Karpathy et al. had layers with up to 512 LSTM cells and were up to 3 layers deep. While the structure of the LSTM cells was the same, the approximately 3 orders of magnitude increase in the network size led to greatly increased performance. The LSTM is similar to the GRU in that it has a number of gates that affect the flow of information (or more accurately based on the order of history, the GRU is similar to the LSTM). The LSTM however has one large distinction from the GRU, it disentangles the hidden state of the cell into two components—(1) the output of the cell, (2) the memory of the cell—whereas the GRU uses the hidden state for both functions.

$$f_t = a_g(W_f \cdot x_t + U_f \cdot h_{t-1}) \tag{8.15}$$

$$i_t = a_g(W_i \cdot x_t + U_i \cdot h_{t-1}) \tag{8.16}$$

$$o_t = a_g(W_o \cdot x_t + U_o \cdot h_{t-1}) \tag{8.17}$$

$$\tilde{c}_t = a_c(W_c \cdot x_t + U_c \cdot h_{t-1}) \tag{8.18}$$

$$c_t = f_t \circ c_{t-1} + i_t \circ \tilde{c}_t \tag{8.19}$$

$$h_t = o_t \circ f_o(c_t) \tag{8.20}$$

> f_t = The forget gate—controls what is saved from the previous step
> i_t = The input gate—controls the update to the memory
> o_t = The output gate—controls what is output from the LSTM cell
> c_t = The memory cell—the memory that is kept additively
> h_t = The hidden output—the output of the LSTM cell

We should see that there are many similarities between the GRU and the LSTM. They both have gates (GRU: z_t, r_t; LSTM: f_t, i_t, o_t) and they both have a component that is a weighted sum of the previous timestep and a new candidate. However, the LSTM has more gates (3) than the GRU (2), and there is a hidden state h_t in addition to the memory cell c_t that seems to function more like the hidden state that we are used to.

The first *major* difference is in the treatment of the previous state. The forget gate is very similar to the GRU's complement of the update gate, when it is large, the previous state (LSTM: c_t; GRU: h_t) is preserved. The input gate therefore acts in a similar way to the GRU's update gate, when it is large, the new hidden state is used. Thus, the LSTM uses two gates where the GRU uses one. The other main difference comes in the decoupling of the previous hidden state and the output. The GRU's hidden state is the same as its output, while the LSTM's are separate.

Both GRUs and LSTMs have seen use in PCGML techniques (GRUs [150], LSTMs [109, 124, 149, 191, 196, 198, 213]), with the only key requirement being that the content can easily be represented as a sequence. The most common usages for GRUs and LSTMs has been level generation for two-dimensional platformer games [149, 150, 191, 196, 198]. These approaches all generate levels one tile at a time in an autoregressive manner. While most of the platformer levels progress in a left-to-right manner, the generation usually happens top-to-bottom-left-to-right (i.e., it generates each column one tile at a time before progressing to the next column). These works have ranged from just generating levels from static levels from a single game [196, 198] to considering blending between multiple games [149, 150] to learning from playtraces derived from videos [191].

The other most common usage has been for rhythm games [66, 109, 213]. Unlike the level generation approaches, most of these approaches are not done in an autoregressive manner. For Dance Dance Gradation [213] the input to the LSTM is another dance chart that is going to be modified based on desired changes to the difficulty level. For the work of Liang et al. [109], the input to the LSTM is audio that the generated beat charts should correspond with. However, the work of Halina and Guzdial [66] uses both audio and previously generated beats to generate in an autoregressive manner for a taiko drumming game.

LSTMs have been used for generating text such as cards for collectible card games [13, 124]. Given that the content is mostly sequential (card text follows the sequential rules for the language it is written in), autoregressive RNNs are a good fit. Furthermore, games such as Magic: The Gathering and Hearthstone have a large corpus of cards (and active fan communities that create custom cards) making training data relatively easy to come by. In the next section, we will work through the card generation task and some of the practical considerations required.

8.3   Sequence-Based Case Study—Card Generation

Returning to card generation, we first need a corpus of cards to learn from for this to be a suitable PCGML task. If we wanted to generate cards for an existing game (like the aforementioned), this would be relatively easy; however, if this is for an in-development game, then things might be harder. We might not have a corpus of sufficient size, and we certainly won't have a corpus the size of something like MtG (over 20,000 at the time of publication). However, for this case-study, we will consider the basics of how to frame card generation as a sequence-based PCGML task.

First, let's consider some cards as seen in Fig. 8.1b. We see two cards "Knight of Procgenmel" and "Keep of Guznodville." While there are obviously differences between them, we see commonalities as well: they both have names, costs, types, text, and stats. Figure 8.1a shows the structure of the cards. All of our cards are going to need to have these fields to be playable, so the first question we come to is, how do we want to order these fields? Our cards need to be represented as a sequence. Within a given field it might be obvious as to how to represent it as a sequence (e.g., for the name and text fields, the content will be represented as a sequence in the language of the card), but we also need to consider how to order the fields themselves.

This might seem like a trivial choice, but it will have further implications. During generation, we will be generating one section at a time, and our generation will be conditioned on the previous content that has been generated. This means that our generator gets to peek backwards as it generates when making a choice. For instance, if the ordering was: ⟨Name, Cost, Type, Text, Attack, Defense⟩, then during the generation of the Type and Text, the Name and Cost would be available as context (e.g., If the Name includes something like "Knight" or "Wizard" it will most likely be a Unit and not a Structure). However, if we want to generate a specific card type, then we would not be able to generate a Name based on that knowledge, because Name comes before Type. In general, this type of generation only works for "mutant shopping" [26]—the generator can produce a bunch of content and the end user can select which ones they like, but there is little to no input from the user (unless the user provides input in the exact order that the generator requires). If we want to allow for any possible combination of these subcomponents to be specified then we will need a different approach—namely the approach discussed in Sect. 8.4.

The next practical consideration is how to represent these cards in a way that can be utilized by our neural network. As in the last chapter, we would want a one-hot encoding, but the question remains: what do the individual dimensions represent? On one end, we could have things represented at the smallest representable unit for the given language (letter for alphabetic languages or characters for ideographic languages). We could also include new characters to represent special symbols (such as attack or defense). On the other end, we could have things represented at a larger level, such as at the word-level for alphabetic languages. There is a trade-off between these two, since the character level is the most general and can represent any possible word, but it requires the model to learn the rules of spelling; on the other hand, the word level won't make spelling mistakes, but it is not possible for words outside the training vocabulary to be generated. In between are **sub-word** approaches that represent common words as words, and uncommon words as sequences of sub-words. E.g., 'the' would likely be its own entry but 'Guznodville' might be split into 'G' 'u' 'z' 'no' 'd' 'ville'. No matter the choice, we will also want special delimiter tokens to represent different fields. For instance, the cards in the above figures might look something like:

```
|Knight of Procgenmel|3|@|When you play ~,
    protect a structure until your next turn|!4|#4|

|Keep of Guznodville|*3%1|&|Your units get +2! and +2#||#8|
```

In this example, special characters were used to represent specific symbols such as '!' for attack and '#' for defense, and a delimiting token of 'I' was utilized to represent the different fields. When training the model, we would train on one card at a time. For each card, we would step through 1 token at a time predicting the next token in the sequence in an autoregressive manner, and we would use Categorical Cross Entropy as the loss. For generation, to generate a card, we would initialize the hidden state of the RNN, and then sample it one token at a time, feeding the sampled token back in as the input. We would stop when we either reach some predefined max limit or an end-of-sequence token. Algorithm 8.1 shows the training setup.

In this example we have used RNN in a generic sense, as there is no difference in the data processing or general training setup between a vanilla RNN, an LSTM, and a GRU. Of course, the fact that the setup is the same between them does not mean that there is no difference between them in terms of performance. While we could use a vanilla RNN, given its limitations it is much more likely that we would want to use either an LSTM or a GRU. While there are differences between GRUs and LSTMs, there is no consistent set of circumstances where one is more desirable over the other. Broadly speaking, while the individual cells of an LSTM are more capable than a GRU, if the same number of trainable parameters is used they are very comparable to each other. In other words, a GRU with approximately 33% more cells than an LSTM will perform roughly the same (they will take up similar amounts of memory and computation time and produce similar results).

8.4   Sequence-to-Sequence Recurrent Neural Networks

In the RNNs we have discussed so far (vanilla, GRU, and LSTM), we have assumed that they are operating in an *autoregressive* fashion—for each input step, we produce an output, which is then fed in as the input for the next step. However, this is not the only input-output modality that is possible for a sequence based approach. Instead of having an output for every input (i.e., one-to-one) as in the autoregressive approaches, we might want a single output for the entire sequence (many-to-one) or multiple outputs for each input (one-to-many). Or, we might not have a direct correspondence between the number of inputs and outputs. Why might we want to do this? Well, first let's consider a classic use case for sequence-to-sequence approaches: translation.

Say we want to translate an utterance from Japanese to English:

"すみません" (sumimasen in Japanese) ⟺ "I'm Sorry"

In Japanese, it is a single word (that depending on the context might mean "I'm Sorry," "Excuse Me," etc.), but in English, it is two words. When translating, it is possible for the utterances to be of different lengths. *We need an approach that can operate on input and output sequences of different sizes.* What we need is an approach that maps from source representation (in this case Japanese, but this could be anything, a level, a character, a rule) to a latent representation and then to a target representation.

Algorithm 8.1 Autoregressive RNN PCGML

1: {**Definitions**}
2: V //Vocabulary of Tokens
3: $v = |V|$ //Size of Vocabulary
4: e //Size of Embedding
5: k //Number of Hidden Layers
6: d //Size of Hidden Layers
7: {**Initialization**}
8: $R \leftarrow \text{RNN}(e, k, d, v)$ //RNN(input dim, layers, hidden size, output dim
9: $E \leftarrow \text{Embedding}(v, e)$
10: {**Training Step**} $- c$ //The current card to learn from
11: $R.\text{reset}()$
12: **for** $tok \in c, next \in \text{shift}(c)$ **do**

13:     $tok_e \leftarrow E[tok]$
14:     $pred \leftarrow RNN(tok_e)$
15:     $p \leftarrow \text{SoftMax}(pred)$
16:     $l \leftarrow L_{CE}(p, next)$ //Loss is Cross Entropy between predictions and truth
17:     backward(l)
18:     step(∇l)

19: **end for**
20: {**Usage Step**}
21: $prev \leftarrow \text{<START>}$
22: $R.\text{reset}()$
23: $gen \leftarrow [\ ]$
24: **for** $i = 1..max$ **do**

25:     $tok_e \leftarrow E[prev]$
26:     $pred \leftarrow RNN(tok_e)$
27:     $p \leftarrow \text{SoftMax}(pred)$
28:     $tok_p \leftarrow \text{Sample}(p)$
29:     $gen.\text{append}(tok_p)$
30:     $prev \leftarrow tok_p$
31:     **if** $tok_p = \text{<END>}$ **then**

32:         break

33:     **end if**

34: **end for**

Enter **sequence-to-sequence** (seq2seq) approaches. In a seq2seq approach, we have two separate recurrent models—an **Encoder** that encodes our input to some latent representation, and a **Decoder** that decodes the representation to the output representation. In many ways this is similar to the AutoEncoders introduced in the previous chapter; However, unlike AutoEncoders we are not trying to recreate the input. Going back to the discussion of input modalities—if instead of having an output at every timestep (one-to-one), we made it so that our encoder only produced a single output (many-to-one), we could use an RNN to make a fixed size latent encoding. We could then use this encoding as the initial hidden state for our

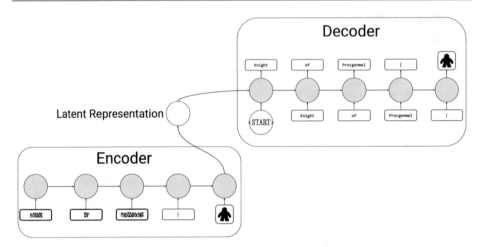

Fig. 8.3 Visual representation of a seq2seq with an Encoder and Decoder. In this setup, the *corrupted* (**bold**) looking tokens are meant to represent tokens that have been removed from the original card. The input, encoding, sequence contains the partial specification, while the decoding sequence contains the full reconstructed card

decoder, and then decode in an autoregressive manner, just as we have previously done. To return to the card generation example, let's assume that we wish to set the type of the card and want to generate everything else. In a purely autoregressive setup we would need to have the card type come first and everything else would follow, but that would lock us in to that specific modality. However, in a seq2seq modality, we can frame this as a translation problem: we want to translate from partial card specifications to fully reified cards. See Fig. 8.3 for a visual representation.

Beyond simply translating between a partial representation to a fully reified one, we can also use these latent representations for new purposes. Let's say that we have lots of levels for a game, and there are few different kinds of levels (e.g., some levels are "normal" while others have lots of floating platforms). If we train our seq2seq model on these different levels, we can then manipulate the encodings to achieve a number of different effects, such as interpolating between pieces of content (discussed below) or finding latent dimensions that correspond to high level features.

Our latent representations are just vectors, and any math that we can perform on vectors, we can also perform on the latent representations. Perhaps simplest, if we have two different encodings we can **interpolate** between them in this latent space—smoothly transitioning from one to the other. We can use this interpolation in latent space to interpolate between two pieces of content. The latent encodings are just vectors and we can manipulate them

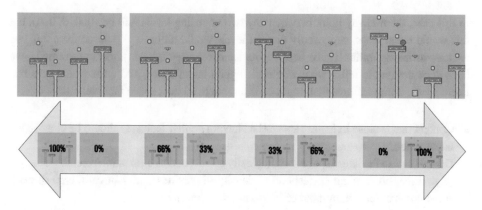

Fig. 8.4 Visual representation of a seq2seq with an Encoder and Decoder

with standard vector math. If we have our encoder—$E()$—, and we have two levels L_A and L_B then we can interpolate between them with $interp(L_A, L_B, i)$ defined as:

$$e_a = E(L_A) \tag{8.21}$$

$$e_b = E(L_B) \tag{8.22}$$

$$d = e_b - e_a \tag{8.23}$$

$$interp(L_A, L_B, i) = e_a + d \times i \tag{8.24}$$

Where i is our interpolant which can take on values between 0 (fully level A) and 1 (fully level B). See Fig. 8.4 for a visual representation of an interpolation. We can then decode this interpolated latent vector to produce a level that is *hopefully* a level between the two original levels. We say "hopefully" because there is no guarantee that the midway point between two levels in interpolation space actually represents the actual midway point between two concrete levels. Imagine a sort of worst case scenario—let's say that we have 10 levels and 5 dimensions. We could imagine a case where the encoder learns to represent each level with a single variable, either -1 or 1 in one of the dimensions. E.g.,

$$E(L_1) = (1, 0, 0, 0, 0) \tag{8.25}$$

$$E(L_2) = (-1, 0, 0, 0, 0) \tag{8.26}$$

$$E(L_3) = (0, 1, 0, 0, 0) \tag{8.27}$$

$$E(L_4) = (0, -1, 0, 0, 0) \tag{8.28}$$

$$\dots \tag{8.29}$$

In this case, the midway point between any two opposing levels will be $(0, 0, 0, 0, 0)$, but this would mean that 5 supposedly distinct levels would all have to emerge from this one vector—an impossibility. This is because our latent representation is under constrained—we have done nothing to tell it what would be a good latent space, and what would be a bad latent space. Some features that might be desirable in a latent space include:

- Content that is close in feature space (i.e., having similar features) should be close in the latent space.
- Dimensions in the latent space should ideally correspond to human interpretable features (e.g., dimensions might correspond to density of enemies, rarity of card, type of card, smoothness of terrain, number of jumps required, etc.).

Some [152] have used **Variational AutoEncoders** (VAEs) [97] to try to achieve these goals. VAEs are a class of AutoEncoder that place additional constraints on the latent space by forcing the AutoEncoder to encode data points as the parameters of a Gaussian Normal distribution—instead of predicting e it predicts e (the mean) and σ (the variance) (i.e., $\mathcal{N}(e, \sigma)$). It then samples from that distribution before decoding the sampled latent representation. Additionally, it adds a penalty to the Loss in the form of the **Kullback-Leibler divergence** (KL divergence) between the encoded distribution and the standard normal distribution (i.e., with mean of 0 and variance of 1, $\mathcal{N}(0, 1)$). KL divergence can be thought of something akin to the distance between two probability distributions, but it is beyond this book to cover in depth [99].

So, why the random sampling and the penalty? The addition of both adds a tension to the model—the model would like to be able to encode things so that it can perfectly reconstruct them in which case it would be able to set the mean e to any value, and to set the variance σ to 0 (i.e., no randomness). However, the KL divergence would penalize this heavily. Conversely, the KL divergence wants to force the encodings to share the same mean ($e = 0$) and to be highly random ($\sigma = 1$), but this would mean that the model would be bad at reconstructing the autoencoded content. This tension means that the differences in latent space most often correspond to differences in content space. If two levels are similar it will try to put their encodings near each other in the latent space, because if the decoding messes up it would not be as bad as it would if they were completely different from each other. However, if two levels are very different, it will try to place them far apart from each other in the latent space, because the reconstruction loss from getting them mixed up would be very high.

Seq2seq models have been popular across a range of content types. Just as with the single sequence RNN approaches mentioned previously, 2D platformer level generation has been used as a domain [151]; however, unlike most of the LSTM and GRU focused approaches in the prior section the seq2seq approaches have covered a wide range of platformer games. All of these approaches have used a VAE modified LSTM or GRU seq2seq Encoder-Decoder architecture to first encode levels in a latent representation, and then used an autoregressive

RNN to generate levels, all with a goal towards **blending** (finding content *between* two known levels) between levels both within a game and between games. Platformers are not the only level type that has been considered as [203] used a VAE-LSTM seq2seq model to generate Angry Birds levels. Furthermore, levels are not the only content that has been handled with a seq2seq approach as [199] used an LSTM seq2seq model to generate Magic: The Gathering cards given a partial specification, just as with the running example from this chapter.

One final note about the seq2seq models—in using a seq2seq model we are basically doubling the length of our sequences (assuming the input and output are of similar sizes). While approaches such as GRUs and LSTMs are capable of learning long sequences, these seq2seq approaches are very taxing on them, and there is a good chance that the model will have forgotten the first token in the input by the time they are trying to decode the corresponding token in the output. Towards this end, an additional mechanism was developed known as **attention** [217], which allows the decoder to look back and "pay attention" to individual elements in the encoding process as opposed to only being given the encoded representation. For generative purposes, attention is not especially useful for seq2seq—as we often want a single latent encoding to manipulate, which is not possible with attention based models. However, a whole class of architectures built solely off of attention have been developed, which can operate on sequences without the need for any recurrent structure— i.e., **the Transformer**. In the next section, we will introduce Transformer architectures, and see what makes them so powerful.

8.5   Transformer Models

To understand how Transformers work, we first need to dig into the idea of attention, and why it is useful in many cases. Figure 8.5 shows a visual representation of how attention works. Instead of encoding the entire input sequence into a single, fixed-length vector, we instead allow the decoder to look back at every single step of the encoding–using a weighted sum of the encodings as part of the input to the decoder (in addition to the autoregressive

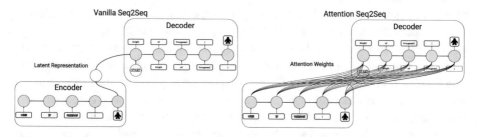

Fig. 8.5 Comparison between a vanilla seq2seq model and a seq2seq model with an attention mechanism

"input" and the decoder's own hidden state)—thereby enabling the decoder to *pay attention* to the input. But our input sequence is of arbitrary size, so how can our decoder learn how to divide its attention? We can't simply use a weight matrix, because we can't a priori know the size of the matrix (and it could change for every sequence).

We need a function that takes in the current state of the decoder D_i and the entire output of the encoder across all timesteps $(E_0...E_n)$ and produces a weighted average of the encoder outputs by selecting how much attention $A(D_i, E_j)$ to apply to each encoding E_j.

$$\text{Attention}(D_i, E_0...E_n) = \sum_{j=0}^{n} A(D_i, E_j)E_j \qquad (8.30)$$

Given that we want a weighted average, we want all $A(D_i, E_j)$ to sum to 1—the easiest way to guarantee this is by the use of the softmax function that we defined in Chp. 7. Now, the only question is what are we taking the softmax of?

The formula for the weight of the attention was originally [9] defined as $A_{add}(D_i, E_j) = v \tanh(W[D_i; E_j])$ where $[a; b]$ means the concatenation of a and b, and v and W are trainable weights. However, following work [217] found that it was possible to achieve good results without the addition of more weights by using **scaled dot product attention** which is defined as $A_{dp}(D_i, E_j) = \frac{D_i \cdot E_j}{\sqrt{d}}$ where d is the dimensionality of D and E. Why the scaling? If D_i and E_i are d dimensional vectors whose components are independent random variables with mean of 0 and variance of 1, then their dot product would have a variance of d—the original intuition is that it would be better if the attention had a variance of 1, hence the scaling by $\sqrt{d}$. So, how can attention be used in a general way for sequences? Well, there's nothing that says that we need a seq2seq architecture to use attention—we could just allow our model to look back at anything from earlier in the sequence, entirely forgoing a recurrent hidden state to carry the previous context. This is called a **causal attention mechanism**, so named because it does not violate causality—it does not "travel through time" and look ahead at the future. Such causal Transformers have, at the time of writing, become de rigueur for text generation with models such as GPT-2 [142] and GPT-3 [19] representing the state of the art.

Say we wanted to make our own G-PCG-PT-2, how would we do it? Well, let's dig into how exactly Attention is used in an autoregressive, causal Transformer. At each layer in the Transformer, there is a node corresponding to each element in the input, whose value is based on causally attending to the previous layer. See Fig. 8.6 for a visual representation.

We see that at the first layer, all nodes are able to pay attention to the first input, but only the final node is able to pay attention to the last input (similarly, the first node is only able to pay attention to the first input). In the second layer, the same patterns remain, except that instead of paying attention to the input, they pay attention to the first hidden layer. Unlike the previous RNN based models, there is no state being carried forward from node to node—and in fact, within a layer there is no dependency on the order that the nodes values

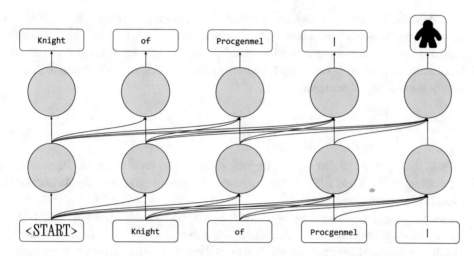

Fig. 8.6 Visual representation of an autoregressive Transformer with a causal attention mask (the attention masked used by GPT-2, GPT-3, and other Transformer Language Models)

are calculated. The only dependencies exist between layers, not within. This might seem trivial, but it is in fact one of the great strengths of Transformer-based approaches—they are more inherently parallelizable than recurrent models (which have dependencies both between and within layers). Formally, our Transformer is defined as:

$$H_{i,k} = W_k \cdot \text{Attention}(H_{i,k-1}, H_{0,k-1}...H_{i,k-1}) \tag{8.31}$$

$$H_{i,0} = x_i \tag{8.32}$$

$$
\boxed{
\begin{aligned}
i &= \text{Timestep} \\
k &= \text{Layer Index} \\
W_k &= \text{The weight matrix for layer } k \\
H_{i,k} &= \text{The output of layer } k \text{ at timestep } i
\end{aligned}
}
$$

Now, one thing may have jumped out as an issue—beyond the fact that some nodes do not get to pay attention to all of the input from the previous layer—there is no notion of ordering. E.g., the second node in the sequence pays attention to the first and second inputs, but it has no way of knowing which is the first and which is the second. With the RNN models we never had any such issue, because the inherent structure of "Process the first input and pass the hidden state to the second node, ..." meant that the RNN knew the ordering of inputs. But Transformers don't have any such implicit ordering, they simply take the weighted average of everything from the previous layer (that they have access to). How then do we modify our Transformer such that it will understand the order of inputs?

Previously, we introduced the concept of embeddings as a way to turn our discrete tiles into vectors of numbers, but there's nothing that says that embeddings are limited to categorical data. Anything that can be represented by an integer, say the timestep of an input, can be used in an embedding layer. In this way, Transformers use a **positional embedding** to directly encode the ordering of the input:

$$H_{i,0} = x_i + p[i] \tag{8.33}$$

That is, the input to the first layer is the sum of the actual input value and its positional embedding. Putting all of this together, we have a causal autoregressive Transformer that is capable of generating sequences.

As of writing, Transformers have become de rigueur for text based generation. GPT-2 is a Large pretrained Language Model (LLM) that was trained on a next token generation task on 40 GB of text scraped from the internet. This gives it access to a host of real world knowledge (and biases) which it can carry over when fine-tuned for a more specific generation task. For instance, Max Woolf used GPT-2 finetuned on Magic:The Gathering cards [221] to generate cards based on real world concepts such as 'The United States of America' (which is a land) and 'Twitter' (which is a goblin). It has also formed the basis for a new form of procedural content generation where it forms the core engine that the player interacts with, as in AI Dungeon 2 [220]. AI Dungeon 2 presents players with a prompt of the form that can be found in class text parser games. AI Dungeon 2 started as GPT-2 fine-tuned on approximately 30 MB of text adventures from https://chooseyourstory.com a community led choose-your-own-adventure website. Players are able to type in freeform text, which allows for the stories to go in very *non-traditional* directions; e.g., a player might recruit a skeleton to join their band, and then go on a tour of the realm, playing shows to sold out crowds [54].

8.5.1   Case Study—Sequence to Sequence Transformer for Card Generation

Let's return to the card generation task. Previously, we defined the task in an autoregressive manner where we first converted the structured fields of the card into a linear sequence, like:

```
|Keep of Guznodville|*3%1|&|Your units get +2! and +2#||#8|
```

However, we ran into the issue that by handling it as a linear sequence we were locked into specifying certain aspects of the card in order. If we want to specify the type of the card (which comes third), we must first specify the name and cost (which come first and second). If instead we treat this as a sequence to sequence approach then we are able to treat the two sequences in an independent order.

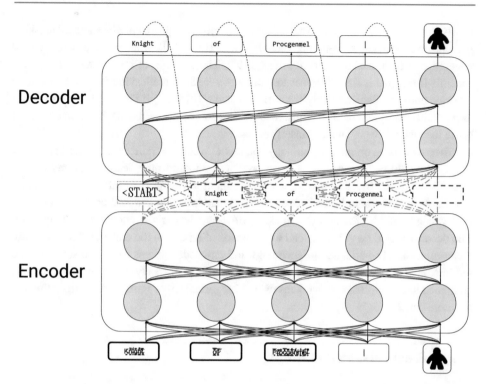

Fig. 8.7 Visual representation of a sequence to sequence Transformer composed of an Encoder with standard full attention and a Decoder with a causal attention mask. While the Decoder has a causal attention mask within the hidden layers and to its token inputs (which are autoregressively fed back in after the initial '<START>'—visualized as blue hyphenated lines), it is able to fully attend to everything from the Encoder (the blue dash-dotted attention lines between the Encoder and Decoder)

For this example, we will utilize two Transformers, one with a full attention mask (the Encoder) and one with a causal attention mask (the Decoder). The Encoder's job is to produce an encoding of the full input, but that full input is going to have **masked** tokens, so the final encoding will only have partial information. The Decoder will operate in an autoregressive manner, predicting the next token one at a time, using the causal attention mask to only pay attention to what it has previously generated (not allowing it to cheat by looking into the future); However, it will not only pay attention to the previously decoded tokens, it is also able to attend to the encoding with a full attention mask. See Fig. 8.7 for a visual representation.

To train such a model, we will need to make two major choices with regards to the masking. First, how many tokens should be masked? Second, how do we want to mask (e.g., token by token, full sections)? For choosing how many tokens to mask, let's consider the extremes. If we only mask 1 token out of a sequence, then it will be a very easy task for the system to fill in the missing piece. Conversely, if we masked all but 1 token it would be a nearly impossible task to reconstruct a card given only a single token. The standard masking percentage used in large pre-trained models is 15% ([36, 117]), while previous work in card generation used a masking rate of 33% ([199]). The question of how to mask is context specific. If the goal is for a designer to input a few fields while leaving the rest blank, then masking entire fields is probably the correct answer. However, if the goal is for a designer to partially enter some fields (for instance writing half of the card text) then it makes more sense to mask tokens or small spans of tokens.

Given a decision on masking strategy, the model would be trained on one card at a time. The card would be used as the target to predict, the input to the decoder (shifted by one after adding a '<START>' token), and as the input to the decoder (after being masked). The masked card would be embedded, a positional embedding would be added, and this embedding would be fed into the encoder which would use self-attention layer by layer. The decoder would be able to attend to the encoded input and the embedded tokens (with a causal mask). The attended to encodings and embedded tokens would be summed and fed into the decoder which would use causal attention layer by layer. This would finally be compared to the original card's tokens using Categorical Cross Entropy. See Algorithm 8.2 for more detail.

8.6   Practical Considerations

One issue when using neural networks is the tension in choosing the size of the model (the number of layers and the number of hidden dimensions). Larger models perform better, but they can also overfit and memorize the training data (especially in data sparse domains such as game content). However, there exist approaches that attempt to minimize overfitting, the most common for sequence based models is **dropout**. Dropout does what it says, it "drops out" some portion of the model randomly at each step of training. Summerville [196] found that increasing the size of the (LSTM-based) model and aggressively using dropout (upwards of 80% dropout) led to improved performance with no memorization as opposed to using a smaller model with no dropout (effectively keeping the same number of parameters at each step of training). It is an open question as to what amount of training data is necessary for a Transformer-based approach to do substantively better than an LSTM-based approach, and if it is even helpful to use a Transformer for a small data domain like PCG.

When encoding a two dimensional representation such as a map in a one dimensional sequence, a decision has to be made as to how this process should be done. There are two particularly obvious ways, left-to-right-top-to-bottom and top-to-bottom-left-to-right (or perhaps bottom-to-top). Given that most levels in side-view games (such as Mario or Megaman) generally have a horizontal orientation (they are wider than they are tall), the better approach is to go top-to-bottom first, as opposed to left-to-right. This is due to the fact that most of the levels are generally only a dozen or so tiles high, while they might be hundreds of tiles wide—which means that if a model is going left-to-right-top-to-bottom then it has to remember across hundreds of timesteps to remember what was just 1 tile above a given tile; conversely, a model going top-to-bottom-left-to-right would only have to remember approximately a dozen timesteps to remember what was 1 tile to the left.

Unlike the Markov Models discussed in Chap. 6, there is no strong preference for going bottom-to-top instead of top-to-bottom in 2D side-view games. This is because the position is either directly (in the case of Transformers) or implicitly (in the case of RNNs) encoded, and a model of sufficient size will be able to remember the position information regardless of the direction. Similarly, there is no strong preference for going in the direction of play (usually left-to-right for side scrolling games)—in fact, some research has shown that there is a slight benefit for going in the reverse direction of play [196]. Of course, none of these matter for games that have an orientation closer to a square—and in those cases the ordering should be shortest-dimension-to-widest-dimension whichever that may be.

The examples in this chapter focused on generating cards, but the same basic approaches would work for a wide variety of content. For 2D content, all of the autoregressive approaches would work, assuming some way to turn the 2D content (e.g., levels, images, icons, etc.) into a linear sequence (such as the above mentioned approaches). For something like the sequence to sequence approach in Sect. 8.5.1, this could be turning a partial 2D specification into a full level (e.g., given a few chunks of a level, fill in the rest). For Transformer approaches with positional embeddings it is also possible to use two embeddings for the x and y positions, as opposed to just using an embedding for the position in the sequence (which implicitly encodes the x and y positions) [134].

8.7 Takeaways

In this chapter we discussed a wide range of sequence-based neural networks, starting from a basic Recurrent Neural Network, through more advanced RNNs (the LSTM and GRU), through different modalities (autoregressive and sequence to sequence), finally coming to Transformer models that are built off of the attention mechanism. These approaches are very well-suited to content that is inherently linear (text, level progressions, storylines, etc.). However, while these are better suited for inherently linear content, they are perfectly capable of handling content that is not necessarily sequential so long as that content can be converted into a sequence. In cases like the running example of card generation, that requires a decision of how to order the given sub-components of the card. For something like 2D level generation this requires more thought and experimentation (although good choices are described in the previous section). In the next chapter we will discuss a class of models that is much more inherently suited to two dimensional content—**Convolutional Neural Networks**.

Algorithm 8.2 Seq2Seq Transformer PCGML

1: {**Definitions**}
2: V //Vocabulary of Tokens
3: $v = |V|$ //Size of Vocabulary
4: s //Maximum supported sequence length
5: e //Size of Embedding
6: k //Number of Hidden Layers
7: m //Masking percentage
8: {**Initialization**}
9: $Enc \leftarrow$ Transformer(e, k, v) //(embedding/hidden dim, layers, output dim)
10: $Dec \leftarrow$ Transformer(e, k, v) //(embedding/hidden dim, layers, output dim)
11: $W_o \leftarrow$ Linear(e, v) //(embedding/hidden dim, layers, output dim)
12: $E \leftarrow$ Embedding(v, e)
13: $P \leftarrow$ Embedding(s, e)
14: {**Training Step**} $- c$ //The current card to learn from
15: $inp \leftarrow$ mask(c, m), $enc \leftarrow [\ ]$
16: **for** $i = 0..$len(c) **do**

17:    enc.append($E[inp[i]] + P[i]$)

18: **end for**
19: **for** $h = 0..k$ **do**

20:    $next \leftarrow []$
21:    **for** $i = 0..$len(c) **do**

22:       $next[i] \leftarrow$ Attention($Enc_{h,i}, enc[0]...enc[$len(c)$]$)

23:    **end for**
24:    $enc \leftarrow next$

25: **end for**
26: $dec \leftarrow [\ ]$
27: **for** $i = 0..$len(c) **do**

28:    dec.append($E[c[i]] + P[i]$)

29: **end for**
30: **for** $h = 0..k$ **do**

31:    $next \leftarrow []$
32:    **for** $i = 0..$len(c) **do**

33:       $next[i] \leftarrow$ Attention($Dec_{h,i}, enc[0]...enc[$len(c)$]$)
34:       $next[i] \leftarrow next[i] +$ Attention($Dec_{h,i}, dec[0]...dec[i]$)

35:    **end for**
36:    $dec \leftarrow next$

37: **end for**
38: **for** $i = 0..$len(c) **do**

39:    $pred \leftarrow W_o \cdot next[i]$
40:    $p \leftarrow$ SoftMax($pred$)
41:    $l \leftarrow L_{CE}(p, c[i])$ //Loss is Cross Entropy between predictions and truth
42:    backward(l)
43:    step(∇l)

44: **end for**

Grid-Based DNN PCGML

9

In the previous chapter we used sequence-based Deep Neural Networks (both Recurrent and Transformer based), but sequences are only one possible way for us to represent the content that we wish to generate. For some types of content (like levels that have one dimension that is much larger than the others), this representation makes sense as the content is engaged sequentially—but it is not the only way for us to represent content. Let's imagine that we have a dungeon type level (as seen in Fig. 9.1). Let's examine this level—we notice some patterns show up (torches on both sides of a door, columns in the corners, the top and bottom of a column are always together) and are repeated throughout the level. We also notice some objects show up in spatial variations of each other (the mushrooms and skeletons). For some of these, there are larger structural patterns (the doors with torches appear at the top and bottom) and for some there aren't (the mushrooms and skeletons can appear wherever). When thinking about how to represent this level, we can imagine that some of these objects—and even larger patterns—don't care about where they are in the level, they only care about a small context around themselves. It's with that lens, that we turn to a class of Deep Neural Network that is focused on "image-like" or "grid-based" data (i.e., has a width and height and each position has a value), the **Convolutional Neural Network** (CNN).

9.1 Convolutions

To discuss CNNs, we must first discuss **convolutions**. In discussing convolutions, we will first introduce what convolutions are (and are not), we will then motivate why convolutions are useful, and then we will discuss CNNs. Convolutions are a way of composing two functions together and are denoted with the $*$ operator and:

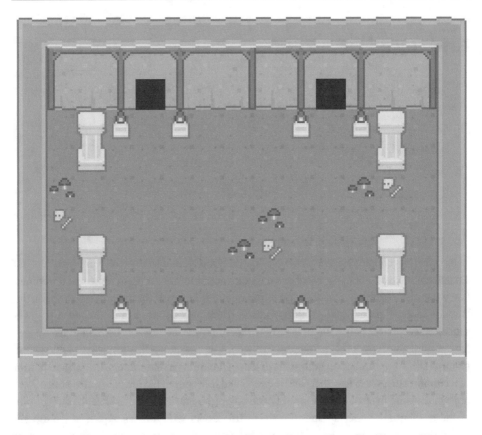

Fig. 9.1 A top-down dungeon level. Assets taken from the Kenney Roguelike Caves and Dungeons 1 Pack

$$(f * g)(n) = \sum_{m=-\infty}^{\infty} f(n - m)g(m) \qquad (9.1)$$

But what does this mean? We can think of this as sliding one function around on top of another to get a new function that is the result of that sliding around. That is, we define a new function $(f * g)(n)$ that is composed of $f(x)$, $g(x)$ where at each point n we sum up the multiplication of the two functions around that point $(f(n - m), g(m))$. Now, these functions might not be functions as you classically think of them, but instead might be *images* and *filters*. Images are something you are most likely familiar with on an intuitive level, but at a computational level they are commonly represented as a 2D matrix of pixels, where each pixel can have some number of channels (i.e., colors—Grey Scale being a single channel, Red-Green-Blue being 3 channels). Thinking of them as a function, an image is a function that takes in the horizontal and vertical positions and the index of the color channel and returns the pixel value for that color at that position: $I(x, y, c) = v$. Filters—at least those applied to images in a convolution—are actually typically represented in an identical manner—a 2D matrix of pixels. The only difference being how we conceive of them—we see an image of a dog and think of it as an image, but a given 3 by 3 subset of an image isn't necessarily recognizable; however, the filter is a function that we want to compose with our image to get a modified version of our original image—some common use cases might be edge detection, blurring, high-pass/low-pass filtering, etc. And while it might not make intuitive sense that images and filters have the exact same representation, this makes sense mathematically—$f * g(n) = g * f(n)$—convolution is commutative, so while we might commonly think of it as applying the filter to the image (e.g., Apply edge detection to an image of a dog), it is mathematically identical to applying the image to the filter (e.g., Apply the dog filter to an image of an edge). While it would be too lengthy to describe the RGB values for an image of a dog, we can think of an edge in an image as when we change colors rapidly in an image. So, for instance, we would have a vertical edge in the grayscale image described by the following matrix:

$$\begin{bmatrix} 0.2 \ 0.2 \ 0.8 \ 0.8 \\ 0.2 \ 0.2 \ 0.8 \ 0.8 \\ 0.2 \ 0.2 \ 0.8 \ 0.8 \\ 0.2 \ 0.2 \ 0.8 \ 0.8 \end{bmatrix} \tag{9.2}$$

Which in greyscale would look like: . In other words, a dark grey region on the left, a light grey region on the right, and a vertical edge in the middle delineating them.

To understand convolutions, we will start with a simple abstract version. To start, we will have 2 matrices:

$$\begin{bmatrix} A & B & C \\ D & E & F \\ G & H & I \end{bmatrix} \tag{9.3}$$

$$\begin{bmatrix} w_{0,0} & w_{1,0} & w_{2,0} \\ w_{0,1} & w_{1,1} & w_{2,1} \\ w_{0,2} & w_{1,2} & w_{2,2} \end{bmatrix} \tag{9.4}$$

There are many different matrix operations we might apply between these two matrices. Before getting into convolutions, let's look at two other common matrix operations, so that we can see how convolution differs. First, let's consider the Hadamard product (element-wise multiplication):

$$\begin{bmatrix} A & B & C \\ D & E & F \\ G & H & I \end{bmatrix} \circ \begin{bmatrix} w_{0,0} & w_{1,0} & w_{2,0} \\ w_{0,1} & w_{1,1} & w_{2,1} \\ w_{0,2} & w_{1,2} & w_{2,2} \end{bmatrix} = \begin{bmatrix} w_{0,0}A & w_{1,0}B & w_{2,0}C \\ w_{0,1}D & w_{1,1}E & w_{2,1}F \\ w_{0,2}G & w_{1,2}H & w_{2,2}I \end{bmatrix} \tag{9.5}$$

Given two matrices (of identical size) this gives us a new matrix where the upper left of one is multiplied by the upper left of the other and so on. This can be useful in certain applications, but it requires two matrices of identical size and each entry in one corresponds to an entry in the other. Convolutions are useful because they (1) do not require identical sizing and (2) the operation uses information from more than just one entry from both matrices. For instance, if we want to detect an edge, we need to know about the values around the edge to see if they change. Moving on, we have matrix multiplication. If we were to perform a matrix multiplication between the above matrices, we would get:

$$\begin{bmatrix} A & B & C \\ D & E & F \\ G & H & I \end{bmatrix} \cdot \begin{bmatrix} w_{0,0} & w_{1,0} & w_{2,0} \\ w_{0,1} & w_{1,1} & w_{2,1} \\ w_{0,2} & w_{1,2} & w_{2,2} \end{bmatrix} =$$

$$\begin{bmatrix} w_{0,0}A + w_{0,1}B + w_{0,2}C & w_{1,0}A + w_{1,1}B + w_{1,2}C & w_{2,0}A + w_{2,1}B + w_{2,2}C \\ w_{0,0}D + w_{0,1}E + w_{0,2}F & w_{1,0}D + w_{1,1}E + w_{1,2}F & w_{2,0}D + w_{2,1}E + w_{2,2}F \\ w_{0,0}G + w_{0,1}H + w_{0,2}I & w_{1,0}G + w_{1,1}H + w_{1,2}I & w_{2,0}G + w_{2,1}H + w_{2,2}I \end{bmatrix} \tag{9.6}$$

To gain an intuition for a matrix multiplication, we multiply the rows of the matrix on the left with the columns of the matrix on the right and then add up the values. For instance, the upper left value of the resulting matrix is made up of the sum of multiplying the first row of the left and the first column of the right. While a matrix multiplication is similar to a convolution in many ways, because they both involve multiplying and adding the values found in matrices, fundamentally they are different operations. The matrix multiplication requires that the matrix on the left hand side's number of columns equals the right hand side's number of rows. For a convolution, the dimensionality of one matrix sets no constraints on the dimensionality of the other. The only impact of the dimensionality of the matrices is in the size of the resulting response matrix. For instance, the convolution of the two matrices defined above would actually only be valid at one point in the matrix—the center. This is because in a convolution we are sliding one matrix around the other, and in this case, the only point where the matrix we are sliding around would fall entirely within the other is when they are lined up with each other. Figure 9.2 shows a visual representation of this convolution. Furthermore, each entry for a matrix multiplication corresponds to one of the rows of the left hand side and one of the columns of the right hand side. Again, convolutions are useful because they operate on all of the entries around a central point. For now, we will just ignore the cells where the filter extends out of bounds, so the convolution would produce:

$$\left[w_{0,0}A + w_{1,0}B + w_{2,0}C + w_{0,1}D + w_{1,1}E + w_{2,1}F + w_{0,2}G + w_{1,2}H + w_{2,2}I\right] \quad (9.7)$$

We notice a number of differences from the matrix multiplications: (1) the resulting matrix is 1×1 not 3×3, (2) rows are multiplied with rows and columns with columns, and (3) each cell corresponds to the summing of the products with the entirety of the matrices (shifted around). Now, we will move on to a more concrete example.

Let's consider the effect of applying a convolution to an image and a filter. Imagine that we have a simple grey-scale image (let's say the values are between 0 and 1) and we have the filter:

$$\begin{bmatrix} 1 & 0 & -1 \\ 1 & 0 & -1 \\ 1 & 0 & -1 \end{bmatrix} \quad (9.8)$$

What would the convolution of image*filter look like? Well, for each location (x, y) in our image we will be using that as the center and then applying our filter at that location—the end result of this being a new image with the filter applied to it. What would our image look like in this case? Well, let's think about under what circumstances this filter would produce a large or small response. We use *response* here as using convolutions to apply filters comes from the traditions of digital signal processing where the response of the filter is convolved with an input signal to produce a third signal.

If our image were filled with the same value (let's say 1) then this would produce a 0 response as we are adding and subtracting the same value 3 times. Conversely, if we had high values on one side (i.e., bright) and low on the other (i.e., dark) then this would sum to a high value. For instance, given the following "image," the value of the center would be:

$$\begin{bmatrix} 0.9 & 0 & 0.01 \\ 0.7 & \mathbf{0} & 0.05 \\ 0.8 & 0 & 0.03 \end{bmatrix} * \begin{bmatrix} 1 & 0 & -1 \\ 1 & 0 & -1 \\ 1 & 0 & -1 \end{bmatrix} =$$

$$0.9 + 0.7 + 0.8 - 0.01 - 0.05 - 0.03 =$$

$$2.31 \quad (9.9)$$

So, this in fact works as an edge detector (this specific filter is commonly referred to as a *Sobel* filter [184])—when we have relatively smooth changes in color and intensity, this produces low values, but when we have a sudden change in color and intensity (i.e., an edge) it produces high responses. See Fig. 9.3 for a visual example.

As a note, Eq. 9.9 shows the response for the center, but a convolution doesn't just calculate the value for a single point, it constructs a new function (in this case, our function can be represented as a matrix). From the point of view of the top-left corner, this would be:

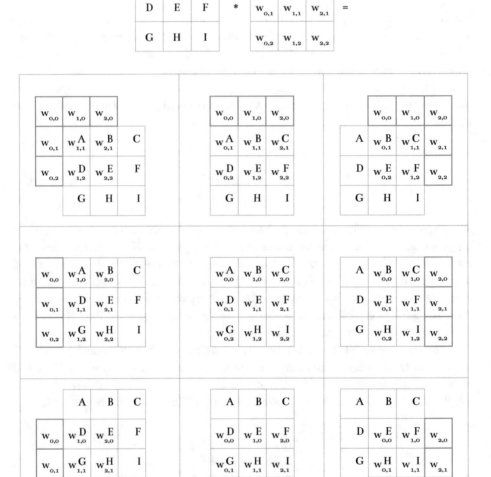

Fig. 9.2 A "visual" representation of a convolution between two matrices. We see that the right matrix is slid around the matrix on the left. Each cell in the resulting matrix would be the sum of the products of the overlapping cells. The red cells denote the cases where the slid matrix extends out of the bounds of the other matrix, and we notice that only in the center do the two matrices perfectly overlap

$$
\begin{bmatrix} \mathbf{0.9} & 0 & 0.01 \\ 0.7 & 0 & 0.05 \\ 0.8 & 0 & 0.03 \end{bmatrix} * \begin{bmatrix} 1 & 0 & -1 \\ 1 & 0 & -1 \\ 1 & 0 & -1 \end{bmatrix} =
$$

$$
1 * OOB + 1 * OOB + 1 * OOB +
$$

$$
0 * OOB + 0 * 0.9 + 0 * 0.7 +
$$

$$
0 * OOB + -1 * OOB + -1 * 0 + -1 * 0 = ??? \tag{9.10}
$$

In this case, the entire left and top portions of the filter are out of bounds (OOB). For now, let's say that anything that falls out of bounds is 0 (other options are discussed in Sect. 9.2). Remember, a convolution is a function that takes in two functions and returns a

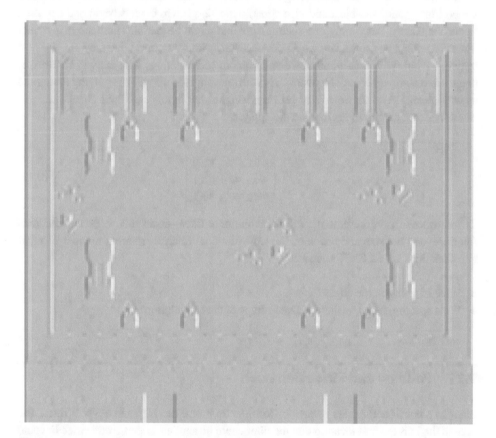

Fig. 9.3 A visualization of the response when applying a vertical edge (Sobel) filter applied to the dungeon image found in Fig. 9.1. Note that the regions where there is a lot of solid color (the grey walls, brown floors) all wind up with a middle (grey) value. The vertical edges that go from dark to light (the left edges of the columns) are dark, while the edges that go from light to dark (the right edges of the columns) are bright

third function (or in our case, takes in two matrices and returns a third matrix). In its entirety, the convolution looks like:

$$\begin{bmatrix} \mathbf{0.9} & 0 & 0.01 \\ 0.7 & 0 & 0.05 \\ 0.8 & 0 & 0.03 \end{bmatrix} * \begin{bmatrix} 1 & 0 & -1 \\ 1 & 0 & -1 \\ 1 & 0 & -1 \end{bmatrix} = \begin{bmatrix} 0 & 1.54 & 0 \\ 0 & 2.31 & 0 \\ 0 & 1.42 & 0 \end{bmatrix} \qquad (9.11)$$

While convolutions are applicable to a host of domains, for our purposes, we can think about them as filters that we are applying to images by sliding them around and doing summation of the products of the matrices. We have seen one example filter so far, the edge detector, and perhaps you question the usefulness of it. Given that edges in an image usually correspond to changes in objects (e.g., in Fig. 9.3 the edges correspond to regions where we go from ground to object and vice versa). Knowing where objects begin and end is an important feature, but is only one-such feature that might be useful. There are many other filters that might be useful, and we do not necessarily know a priori what they are, nor how to define them. With that in mind, we will combine convolutions with neural networks, to create *Convolutional* Neural Networks. As with convolutions we will be applying filters, and as with neural networks we will be learning weights, but in this case the trainable weights of our network are the filters that we are applying.

$$W_k = \begin{bmatrix} w_{0,0} & w_{1,0} & w_{2,0} \\ w_{0,1} & w_{1,1} & w_{2,1} \\ w_{0,2} & w_{1,2} & w_{2,2} \end{bmatrix} \qquad (9.12)$$

The above matrix represents the kth Convolutional filter—itself a 3×3 Convolutional filter. Beyond the size and shape of the Convolutional filter, there are a number of other considerations with a CNN, such as:

- What happens on the edges?
- What spacing, if any, is there between steps of the convolution?

9.2   Padding and Stride Behavior

First, let's consider the edge behavior. Consider an 8×8 tile map and a 3×3 filter. The second tile in is the first location that the filter could be applied, because otherwise it would extend off the edge of the image. See Fig. 9.4 for a visual representation. However, this would mean that applying the filter would reduce the size of the image by two (one from the left/top and one from the right/bottom). If we add some **padding** (extra tiles around our

Fig. 9.4 An 8× 8 tile map with a 3 × 3 filter

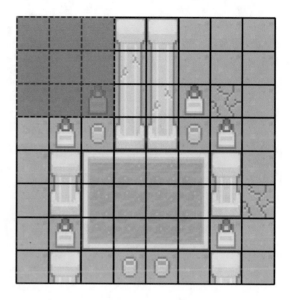

input), we can keep the output of the convolution the same size as the input. While we might want to change the size of our image, we want it to be under our control, not just as a side effect of applying a filter.

The simplest option is to use a **Constant** (most commonly 0) value. This has some benefits (all 0's means that the filters can easily learn what the edges of the image are like) and some drawbacks (all 0's are extremely unlike the rest of the image). Beyond that, the other options all utilize the image itself to derive the padding values. See Fig. 9.5 for a visual representation. **Circular** padding treats the image as if it were a circle—going off one end just wraps you around to the other side. This can also be thought of as the image just endlessly repeating. **Mirror** padding reflects the original image, acting as if it was reversed. **Replicated** padding is a combination of the two—it reverses the original image—like Mirror padding—but also endlessly tiles it—like Circular padding.

Beyond the padding, there is also spacing between applications of the filters to consider. In the original mathematical formulation of a convolution, there is no spacing, but this is often done in CNNs. This spacing—or **Stride**—is a parameter that determines what (if any) gaps exist between applications of the filter. With a stride of 1 we will overlap by 2 tiles every step (see Fig. 9.6a). If we instead had a stride of 3, then we still have full coverage, but with no overlap (see Fig. 9.6b). We can see that 4 steps of the 3 stride Convolutional filter cover more of the input than the 6 steps of the 1 stride filter, and thus would require less computation. Requiring less computation is a desirable feature; however, there are drawbacks to using a larger stride. Namely, there are patterns that will be missed by the filters. In the example in the figure, the section with 2 columns side-by-side will only be seen justified to the left, never justified to the right. This might not be an issue, and might be worth the computational

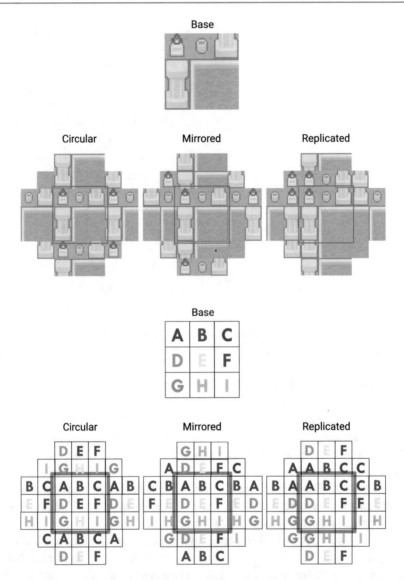

Fig. 9.5 Visual depictions of the main forms of paddings—Circular (which loops around), Mirrored (which reflects at the edges), and Replicated (which reflects and circles around)

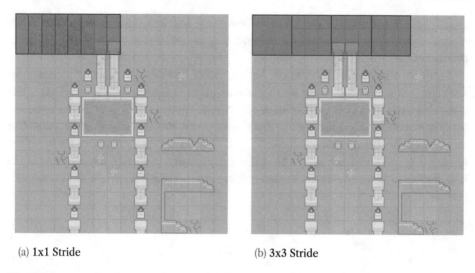

(a) 1x1 Stride   (b) 3x3 Stride

Fig. 9.6 A comparison between the same size filter (3×3) with a stride of 1 (**a**) and a stride of 3 (**b**)

efficiency, but it is still a point in favor of the smaller stride. Generally, in a CNN a mixture of strides will be used, most commonly a number of layers will exist at a given size, and then there will be some form of condensing (e.g., by using a larger stride) that reduces the size—with this pattern repeating. Typically, this pattern will also utilize **Residual** layers. A residual layer is one in which the input is both passed through a Convolutional layer and fed directly into the next layer—see Fig. 9.7 for a visual representation.

What purpose does a residual connection have? It seems somewhat unintuitive that we would directly pass an input directly to the next layer—if that's what we want to do, then what are the layers themselves doing? The answer lies in the previous chapter—it's all about the flow of the gradient and combating the vanishing gradient problem. The residual connections allow the gradient to freely flow backwards through the network, just as the

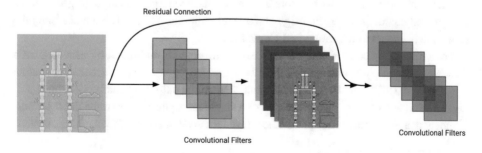

Fig. 9.7 Visual representation of a **Residual** connection in an CNN

mechanisms from the GRUs and LSTMs allowed it to flow backwards through time. While residual connections are helpful for training, they do nothing to combat the problem of overfitting.

There are a number of ways to combat overfitting, but the most popular approach for Convolutional networks is to use **batch normalization**. Batch normalization is a layer that is inserted into CNNs after the non-linear activation is applied to a Convolutional layer. The batch normalization layer modifies its input based on the statistics of the current *batch* (the set of datapoints being considered, recall Chap. 7), by *normalizing* it so that the batch has a mean of 0 and a variance of 1. This normalization helps to smooth the gradients of the loss, leading to faster convergence when training [147]. The equation below describes this process:

$$x_t, ..., x_{t+b} \tag{9.13}$$

$$\mu_t = \frac{1}{b} \sum_{i=t}^{t+b} x_i \tag{9.14}$$

$$\sigma_t = \frac{1}{b} \sum_{i=t}^{t+b} (x_i - \mu_t)^2 \tag{9.15}$$

$$BN_i = \frac{x_i - \mu_t}{\sqrt{\sigma_t}} \tag{9.16}$$

$$
\begin{aligned}
x_t, ..., x_{t+b} &= \text{Output of neuron for a batch} \\
\mu_t &= \text{Mean of output of neuron for a batch} \\
\sigma_t &= \text{Variance of output of neuron for a batch} \\
BN_i &= \text{Normalized output for datapoint } i \text{ in the batch}
\end{aligned}
$$

Now that we know how to train a CNN, we move on to how they can be used for generation. For this we will return to a previous concept, and make use of an AutoEncoder. In a Convolutional AutoEncoder, we use encoding layers that downsample (i.e., the strided convolutions) to wind up with a compressed latent encoding. We can then go from this compressed encoding back to the original size by upsampling. The most common upsampling technique is known as a **Transposed Convolution**.[1]

For a transposed convolution, each element of the input is multiplied by each element of the filter and multiple applications of the filter are added together to produce the final output. E.g., Given a 2×2 filter, F, and a 2×2 input, I, then the output, O can be visualized as in Fig. 9.8. The outer corners are composed of a single multiplication between the outer edge of the input and the outer edge of the filter, the edges are composed of the multiplications of

---

[1] This is sometimes incorrectly referred to as a "deconvolution" but a deconvolution tries to produce two functions from an input function, such that the two deconvolved functions produce the the input function when convolved.

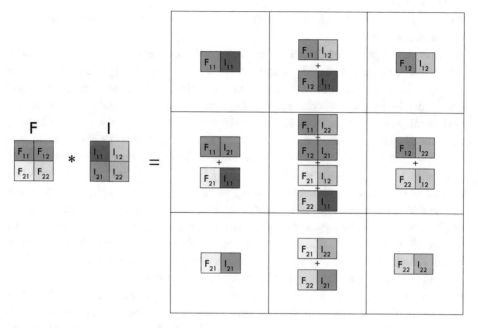

Fig. 9.8 Visual representation of a transposed convolution

the two edges, and the interior is composed of all of the elements. This results in an output that is larger than the input, allowing for the Decoder to upsample. As a note, each layer is also usually made up of a number of filter **channels** (i.e., individual filters). The number of channels in the first layer of the Encoder would correspond to the number of input channels (e.g., in the case of a standard image there would be 3 channels—red, blue, green). The number of channels in the hidden layer is something that would be up to experimentation—but commonly there are between 16 and 512 (usually some factor of 2) channels per hidden layer. Finally, the number of channels in the output would be the same as the input for the AutoEncoder case.

In the case of something like the AutoEncoder we try to recreate the input via an encoding and decoding, with the hope that the compression leads to good generalization—but there is nothing that says that the input and output need to be the same. We can use mismatched inputs and outputs to achieve different effects. For instance, we might already have the coarse geometry of a level (perhaps generated by some other approach) and now we want to place enemies and resources (this is the approach taken by [107]). Instead of using an AutoEncoder to go from Level → Level we instead train a network to go from Level → Resources. The setup for this would be similar to the AutoEncoder in that our input and output will be of the same size. However, we would use different channels for the input and output. E.g., if our data had channels for level geometry, enemies, and resources, we would just use the level geometry channel in our input and just the enemy and resource channels in the output. This is in contrast to the Autoencoder that would use all three channels for both input and output.

CNNs have been a popular choice for PCG, with level generation being a common task. In work by Jain et al. [83] AutoEncoders were used to repair broken content. By training a denoising AutoEncoder, content that was close to functional was encoded and then decoded—with this process repairing the content such that it would have the desired mechanical properties (i.e., in a platformer where there was previously a wall that would block the player's progress, the wall might be shortened such that the player could now jump over it). A more traditional use of AutoEncoders for level generation has been used by Sarkar et al. [152] and Snodgrass and Sarkar [183], both of which used convolutional VAEs to generate platformer levels across a variety of games. While not explicitly framed as a PCG problem, Earle [39] used CNNs with shared weights to play SimCity, which involves the placement of different types of structures which can be viewed as PCG given that many PCG problems involve the placement of content. Chen et al. [24] used CNNs to convert arbitrary images into levels, by trying to predict the tile type given a section of pixels.

Level generation is not the only PCG task that people have applied CNNs to, as sprite and character generation has also been a popular domain. Gonzalez et al. [51] used a VAE that could be conditioned on a number of different words to generate silhouettes of characters. Similar work by Gonzalez et al. [50] used a conditional VAE to recolor sprites based on changes to their attributes (e.g., a red and orange sprite associated with fire might be recolored with white and blue when associated with ice).

The original use of AutoEncoders was to be able to pretrain a model so that the internal representations would be useful for some later task. The idea being that it was easy to get lots of unlabeled data to train an AutoEncoder with, and then the trained AutoEncoder could be fine-tuned on harder to find labeled data. While PCG researchers and practitioners have been able to use AutoEncoders for PCG, there exists a class of Convolutional Neural Networks that are solely devoted to generating new content, **Generative Adversarial Networks**.

9.3   Generative Adversarial Networks

Generative Adversarial Networks (GANs) [52] are a technique that reimagines the standard training process. Instead of pairs of input and output—which it should be stated in our case were often the same—we instead only have exemplar artifacts of the kind that we would wish to generate. In previously discussed methods (Auto-regressive Markov sequence approaches, AutoEncoders, etc.), we trained a model with the task of trying to recreate the examples (by predicting the next item in the sequence or by trying to directly recreate the original after some compression). However, GANs operate in a different manner. Instead of trying to recreate a specific instance of content, they try to generate content that could plausibly pass as the original content. GANs are broken up into two separate networks that compete with each other—the **Generator** and **Discriminator**. The aptly named **Generator** ($\mathcal{G}(z)$)

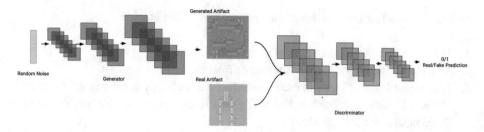

Fig. 9.9 Visual example of a **Generative Adversarial Network** (GAN) with Generator and Discriminator

takes in random noise (z) and returns a piece of content. Unlike the previous approaches which then scored the generated content by how closely it specifically matches a piece of original content, there is a **Discriminator** ($\mathcal{D}(x)$) which takes in a piece of content (x) and returns the probability that content came from the original dataset. A visual representation of a GAN is shown in Fig. 9.9. This is interesting for a number of reasons.

We are no longer trying to directly recreate instances from our training set. We only care about the Generator being able to create artifacts that can be passed off as real (i.e., imagine someone with a forgery saying they had a new work by a famous artist, there is no original to compare it to, but they want people to believe that it isn't a fake). We have two networks working in opposition—but this is something like a friendly opposition (i.e., imagine splitting a sports team in half to scrimmage—each side is trying to win, but at the end of the day it is an exercise that helps both halves learn and improve).

Instead of calculating the loss between a generated piece of content and a piece of content in the training data, GANs have two components to their loss. (1) Can the Discriminator successfully tell true (original) content from false (generated)? (2) Can the Generator successfully trick the Discriminator? This means that the GAN can make subtle distinctions that weren't previously possible. To give an example, imagine a checkerboard pattern where each pixel alternates white and black—if we were to shift it by 1 pixel to the right, our standard losses would treat this as extremely bad as we now predict black when it should be white, and vice versa. However, perceptually, these are very similar, and it is possible that a GAN might learn that both patterns could be used interchangeably.

So, how does this actually work? The training takes place in a few steps:

1. The Generator, $\mathcal{G}$, produces a set of generated artifacts—$A_{\mathcal{G}}$
2. An equal number of real artifacts are taken from the training data—$A_{\mathcal{R}}$
3. The Discriminator, $\mathcal{D}$, tries to predict which of the artifacts are real and which are fake. The loss (which is defined in Eqs. 9.17–9.19) is then passed backwards through the Discriminator and it is updated
4. After a number of iterations of 1–3, the loss is passed backwards through the Discriminator to the Generator almost identically to step 3 (Eq. 9.20), except in the opposite direction as the Generator wants the Discriminator to incorrectly identify its output as real (Eq. 9.21)
5. Steps 1–4 are repeated until training has converged

The **Discriminator** is given both real artifacts (the training data) and generated ones and produces a prediction for the probability that its input was real. This prediction is scored with a Binary Cross Entropy Loss:

$$\mathcal{D}(\text{artifact}) = P(\text{real}) = p \qquad (9.17)$$

$$y = \begin{cases} 0 & \text{artifact is fake} \\ 1 & \text{artifact is real} \end{cases} \qquad (9.18)$$

$$L_{BCE} = -(y \log(p) + (1 - y) \log(1 - p)) \qquad (9.19)$$

The goal for the Discriminator is thus to minimize its error in predicting real vs. fake:

$$L_{BCE}(\mathcal{D}) = \sum_{a_g \in A_{\mathcal{G}}, a_r \in A_{\mathcal{R}}} -\log(\mathcal{D}(a_r)) - \log(1 - \mathcal{D}(\mathcal{G}(a_g))) \qquad (9.20)$$

However, the goal for the Generator is to try to trick the Discriminator into predicting that it is producing real artifacts—thus it wants to maximize the probability that the generated artifacts are predicted as real. This means it wants to minimize the probability that it is predicted as fake (the second log term in the above equation)—this means the removal of the negation of the log.

$$L_{BCE}(\mathcal{G}) = \sum_{a_g \in A_{\mathcal{G}}} \log(1 - \mathcal{D}(\mathcal{G}(a_g))) \qquad (9.21)$$

To unpack, if the Discriminator predicts that the Generator's output is generated (e.g., it predicts a 0), then this is $\log(1) = 0$; however, if the Discriminator predicts that the Generator's output is real (e.g., it predicts a 1), then this is $\log(0) = -\infty$. Since the Generator wants the Discriminator to predict that its output is real, it is trying to minimize this loss. This might seem odd, but it is important to note that when these losses are being applied, we are only updating one of the networks at a time. Thus, when we are minimizing the loss for the Discriminator, we are only updating the Discriminator—we aren't improving its loss by making the Generator worse. Conversely, when we are minimizing the loss for the Generator, we are only updating the Generator—we aren't improving its loss by making the Discriminator worse. In this way, the networks work adversarially to improve each other. As one of them gets better, the other must also get better to maintain parity. The fact that this whole process is differentiable means that the gradients from the Discriminator flow all the way back to the Generator—which means that the Generator gets information on how to improve by directly getting information on why the Discriminator predicted it as real vs. fake.

The Generator portion of the GAN is usually a fully Convolutional network that starts from some random noise which is reshaped to the correct dimensions such that after a number of upsampling Convolutional layers it will produce an output of the same size and dimensions as the originals. To generate new artifacts, random noise is supplied to the generator which processes it to produce a generated artifact.

From a practical standpoint GANs are difficult to train. In part, there is some inherent difficulty in that it is no longer the case that we can track the loss of the model to determine how the training is proceeding. This is because we have two losses that are adversarial, we should expect that as one goes down, the other goes up, and vice versa. This will result in the loss values bouncing around constantly, never converging. Usually, it requires human inspection to determine if the content is being generated satisfactorily, as the loss values alone can not tell whether the GAN is performing as desired.

Unlike any of the AutoEncoders that we have discussed, we do not get the latent encoding of an artifact—because we do not have an Encoder! Instead, we have a Generator that works in a similar way to the Decoder except instead of producing an artifact (that itself should be a reproduction of the input to the Encoder) from an encoding, it produces an artifact from random noise. But we might want to do something like the interpolation described in Chap. 8. To do this, we need to run our Generator in the reverse direction, taking an artifact and trying to reverse-engineer the random noise that would have produced it. To do so, we take advantage of Stochastic Gradient Descent! Previously, we used Stochastic Gradient Descent to determine how to change the weights of the network to better produce the output that we want given the input. However, if we **freeze** the weights of the network and instead let it change the input itself, we can use Stochastic Gradient Descent to change the input to better replicate the output. This process is not without problems, and it is often the case that a network cannot perfectly replicate an artifact—see Fig. 9.10 for a visual representation. Nonetheless, it enables us to perform manipulations of the input—such as interpolation or vector analogies (e.g., King-Man + Woman = Queen).

Perhaps it is unsurprising that an architecture that is focused on generation has been popular for procedural content generation, but a wide range of PCG domains have been

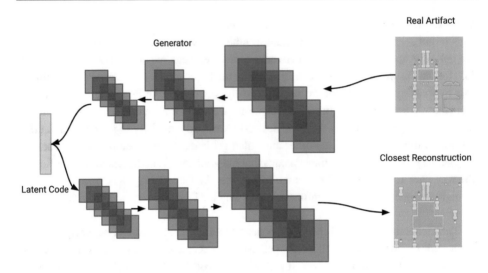

Fig. 9.10 An example of the discrepancy between an artifact and the closest reproduction that is representable by a GAN

tackled via GANs. Once again, level generation has been popular, with platformer level generation still being the most common. Volz et al. [219] trained a GAN on level segments from Super Mario Bros. and then used genetic algorithms to optimize levels based on the latent encodings. This was extended by Schrum et al. [154] to incorporate the evolution into an interactive mixed initiative approach (more of which will be discussed in Chap. 11). This was followed by Capps and Schrum [23] who trained multiple GANs on level segments from Mega Man with the different GANs trained to produce level segments based on the player's movement. The aforementioned approaches all generate fixed size patches of content (usually 16×16) and then stitch them together to create larger levels, but the work of [7] uses a SinGAN that operates at various scales to generate variable sized content. Other types of level content have been considered, with 2D dungeons [154], first-person shooter levels [48], and open world maps [8] as examples.

Furthermore, while levels have been a particularly popular domain for PCG via GANs, they have also been applied to other types of content. Horsley and Perez-Liebana [79] used deep convolutional GANs to generate sprite content for games. This was the inspiration for Henrichs et al. [73] who used AutoEncoders as a pretraining step for the GAN to speed the training process. Both of these approaches generated images as 24-bit RGB images, but Lazarou [105] used a fixed size palette (16 colors in this case), an approach common amongst pixel art practitioners (stemming from historical hardware limitations). Instead of producing an intensity for RGB values in a 0–1 range, it instead uses a softmax activation to predict 1 of 16 colors from a palette that is co-generated with the image, resulting in a much cleaner image with less noise than the previous approaches.

Convolutional approaches, both those based around AutoEncoders and more recently from GANs, are suited to a wide range of PCG tasks. However, there are scenarios where they are not particularly well suited, which the following section will discuss.

9.4 Practical Considerations

Convolutional models have a number of desirable features, but these come with some caveats. Convolutional approaches, like the Recurrent approaches discussed in Chap. 8, can handle arbitrary sized input. So long as the model is fully Convolutional (i.e., this is not true for models such as GANs or VAEs that utilize non-Convolutional layers), none of the layers make any assumption about the size, shape, or orientation of the input. However, unlike the Recurrent models, a model trained on one size/shape is likely going to do extremely poorly for other sizes and shapes. To be fair, there are many cases where Recurrent models will also do poorly (i.e., train on square input and try to operate on drastically different aspect ratios), but there are also cases where Recurrent models will do well where a Convolutional model will not. For instance, in the case of platformer levels, they often have a fixed height but might have arbitrary width. A Recurrent model will learn this (height is fixed, width is variable), but a Convolutional approach will be unequipped to handle different sized input. In part, this is because the scale and scope of what is learned in a Convolutional model is directly tied to how deep it is in the model. For instance, the first layer operates directly on the input, and the second operates on the output of the first, and so on—if the first layer learns patterns like edge detectors, the second layer detects patterns in these edges, and so on—with each layer learning higher abstractions. Often, the number of layers is directly tied to the input size—with the final layers having access to information from across the full scope of the input. If the size/shape changes, then this is no longer the case, and the model will not operate in the way in which it was trained.

Just as with the Recurrent models vs. Transformer based models, dataset sizes are an issue when considering Convolutional approaches. For instance, GANs are very data hungry, and often require large datasets (usually 10,000+ training instances) due to the fact that the Discriminator overfits and completely memorizes the training data for small datasets. This can be counteracted by reducing the capacity of the Discriminator (i.e., making it smaller), but this comes at a price. By making the model smaller, the model is indeed less capable of memorizing the training data, but it is also less capable in general, and will not be able to perform as well as a larger model trained on more data. Because of this, it is often necessary to artificially augment the amount of data. The manipulations that are applied depend on the use case: e.g., for images involving natural subject matters (flowers, dogs, etc.) one can often mirror the data horizontally but not vertically (the sky should not be at the bottom of the image), but for images involving text mirroring will not produce valid input data—for some content it might be acceptable to mirror vertically and horizontally (e.g., the Dungeon levels in this chapter), while it might be unacceptable for others (e.g., the flow of a platformer level might assume left-to-right progression). In the next section, we will consider some PCGML use cases using different convolutional techniques.

9.5   Case Study—CNN Variational Autoencoder for Level Generation

Let's return to the AutoEncoder case study from Chap. 7, but this time we are going to make it a convolutional Variational AutoEncoder (VAE). Again, this will be a model with an **Encoder** (data → latent) and **Decoder** (latent → data); However, there are going to be two major differences between our model here and the previous example.

1. We will use Convolutional `conv2d` layers instead of fully connected `Linear` layers
2. We will apply random noise to the latent representation with an associated penalty

To address the first, the major difference from before comes with the interaction with the latent representation. The convolutional layers do not care about the size of the input so long as the number of channels match up. For example, a convolutional layer with **kernel** (aka filter) size of 3 and 3 input channels can work on a 512×512 RGB image or a 32×32 RGB image, the only difference is that the output size will be different. This means that the size of the output from a convolutional layer is a function of both its input and its parameters. For a convolutional filter with k kernel size, s stride, and p padding, the output width and height ($w_o \times h_o$) for an input image with width and height ($w_i \times h_i$) will be:

$$w_o = \lfloor w_i - k + 2p \rfloor / s + 1 \tag{9.22}$$

$$h_o = \lfloor h_i - k + 2p \rfloor / s + 1 \tag{9.23}$$

While the convolutional layers do not care about the size of their input, the latent layer sizes need to be calculated for the chosen training data dimensionality.

   To address the second, this is just as discussed in Sect. 8.4. The latent encoding is used to produce the mean μ and variance σ for a Gaussian Normal $N(\mu, \sigma)$ distribution. This distribution is then sampled to determine the latent representation that will be passed to the decoder during training. This acts as a regularization technique to produce a robust representation where small changes in the latent encoding should produce small changes in the produced output. To prevent the model from producing a variance of 0 (i.e., no random noise) it is penalized as the produced distribution varies from a Gaussian Normal distribution with a mean of 0 and variance of 1. This forces the model to keep its representations centered around 0 (the mean) and trade off between accuracy and robustness (the variance). Algorithm 9.1 shows pseudocode for this model, and Fig. 9.11 shows a visual representation. From a PCG standpoint, this allows for more control over the generated content.

Algorithm 9.1 VAE CNN

1: {**Definitions**}
2: V //Vocabulary of Tile Types
3: $v = |V|$ //Size of Vocabulary
4: l //Number of Hidden Layers for Encoder and Decoder
5: K //List of kernel sizes
6: C_i //List of input channels
7: C_o ///List of output channels
8: a //Hidden Activation
9: z_i //Input to the Latent size
10: z //Latent size
11: {**Initialization**}
12: $E_{i=0}^{i=l} \leftarrow$ conv2d($K[i], C_i[i], C_o[i]$) //Encoder
13: $D_{i=0}^{i=l} \leftarrow$ conv2d($K[l-i], C_o[l-i], C_i[l-i]$) //Decoder (mirror of Encoder)
14: $W_\mu, W_\sigma = $ Linear(z_i, z)
15: {**Training Step**} $- I$ //The "image" to predict
16: $x \leftarrow I$
17: **for** $i = 0..l$ **do**
18:    $x \leftarrow a(E_i(x))$
19: **end for**
20: $\mu \leftarrow x$.flatten() $\cdot W_\mu$
21: $\sigma \leftarrow x$.flatten() $\cdot W_\sigma$
22: $x \leftarrow N(\mu, \sigma)$
23: **for** $i = 0..l$ **do**
24:    $x \leftarrow a(D_i(x))$
25: **end for**
26: $p \leftarrow$ SoftMax(x)
27: $L \leftarrow L_{CE}(p, I)$ //Loss is Cross Entropy between predictions and truth
28: $L \leftarrow$ KL-div($N(\mu, \sigma), N(0, 1)$) //Loss also includes Variational term
29: backward(L)
30: step(∇L)

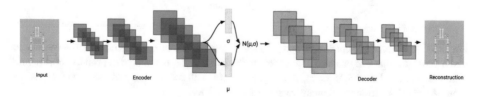

Fig. 9.11 A visual representation of a convolutional Variational AutoEncoder

This specific model is intended to work in a way similar to the previous AutoEncoder which had a categorical tile representation (hence the Softmax activation on the output and the Categorical Cross Entropy loss); However, a very similar model could be applied directly to image data with the major difference being the handling of the output. In that case, there would be no activation function on the output and the loss would be Mean Squared Error (or possibly the Huber loss which is a cross between Mean Squared Error and Mean Absolute Error that has a gentler handling of outliers).

9.6   Case Study—GANs for Sprite Generation

Let's now turn to a different content type, sprites. Figure 9.12 provides a visual example of our goal. Given initial sprites we would like a generator that can produce new sprites with a similar style, but is able to produce sprites with different skin and hair color, poses, clothing, etc. that is still recognizably of a similar style as the input.

To do so, we will represent the sprites as RGB images. The Generator aspect of the GAN will take in random noise and produce RGB images. The Discriminator will take in RGB images and return the probability that the image is real (as opposed to generated). The Generator will be trained by trying to maximize the probability that its content is predicted to be real, while the Discriminator will be given a balanced set of real and fake images and will be trained to correctly classify them.

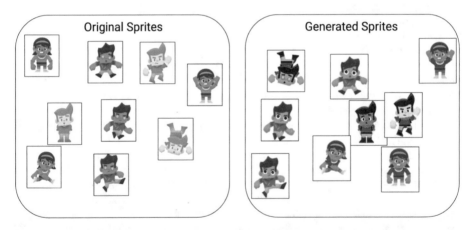

Fig. 9.12 An example of the desired end product for sprite generation. Given initial sprites we would like a generator that can produce new sprites with similar style, but is able to produce sprites with different skin and hair color, poses, clothing, etc. that is still recognizably of the same style as the input. Assets taken from the Kenney Toon Characters 1 Pack

---

Algorithm 9.2 GAN Sprite Generator

1: {**Definitions**}
2: c //Size of the palette
3: l //Number of Hidden Layers for Generator and Discriminator
4: K //List of kernel sizes
5: C_i //List of input channels
6: C_o ///List of output channels
7: a //Hidden Activation
8: z //Random noise size
9: z_g, z_d //Size of generator/discriminator output
10: {**Initialization**}
11: $G_{i=0}^{i=l} \leftarrow \text{conv2d}(K[i], C_i[i], C_o[i])$ //Generator
12: $P \leftarrow \text{Linear}(z_g, c \cdot 3)$ //Palette Weight
13: $I \leftarrow \text{Linear}(z_g, z_g \cdot c)$ //Palette Index
14: $D_{i=0}^{i=l} \leftarrow \text{conv2d}(K[l-i], C_o[l-i], C_i[l-i])$ //Discriminator
15: $O \leftarrow \text{Linear}(z_d, 1)$
16: {**Training Step – Generator**}
17: $x \leftarrow \text{random}(z)$
18: **for** $i = 0..l$ **do**
19:     $x \leftarrow a(G_i(x))$
20: **end for**
21: $p \leftarrow x \cdot P$ //Generate palette
22: $x \leftarrow x \cdot I$ //Generate indices
23: $x \leftarrow \text{SoftMax}(x) \cdot p$ //Use palette to color image
24: $tf \leftarrow \text{Discriminator}(x)$
25: $L \leftarrow L_{BCE}(tf, 1)$
26: backward(L)
27: step(∇L)
28: {**Training Step – Discriminator**}
29: $X \leftarrow [(\text{sample}(dataset, n), 1)] + [(\text{Generator}(\text{random}(z)), 0) \times n]$
30: shuffle(X)
31: **for** $x, y \in X$ **do**
32:     **for** $i = 0..l$ **do**
33:         $x \leftarrow a(D_i(x))$
34:     **end for**
35:     $tf \leftarrow x \cdot O$
36:     $tf \leftarrow \text{Discriminator}(x)$
37:     $L \leftarrow L_{BCE}(tf, y)$
38:     backward(L)
39:     step(∇L)
40: **end for**

We will also take inspiration from Lazarou [105] and the generator will have an intermediate step of generating a palette and a palettized sprite. The choice to do so is two-fold: (1) the output images will be "cleaner" in that they will have a limited set of colors which will avoid a smudgy, blurry look, and (2) artist's often use a similar constraint, which requires visual consistency across and within content. In this case, we will utilize 11 colors: primary skin color, secondary skin color, sclera color, iris color, primary shirt color, secondary shirt color, accessory color, primary shoe color, secondary shoe color, primary pant color, and secondary pant color. The palettized sprite will have a softmax activation and then the palette colors will be multiplied by the palette probabilities to produce the final RGB sprite. Algorithm 9.2 provides pseudocode for this setup.

For other forms of content, this palettization step could be unnecessary, particularly for datasets where any given image contains hundreds or thousands, if not millions, of colors. In that case, the palettization step would instead involve producing just 3 output channels for RGB. On the other hand, for something like level content it would still have a softmax activation to generate tile indices (like the generation of palette indices), but there would be no palettization step. Once we've trained our GAN to our satisfaction, we could use the output sprites directly in a game, use them as the basis of sprites as in Serpa and Rodrigues [156], or even use the trained generator itself within the game.

9.7   Takeaways

In this chapter we discussed Convolutional Neural Networks, a class of Neural Networks that operate on arbitrary-sized input via the use of convolutional operations (i.e., applying filters in a sliding window around an input). These approaches are well-suited to data that is of a fixed-size and aspect-ratio. For the most common generation use cases, an AutoEncoder (variational or vanilla) is likely to be the best option due to its capabilities and ability to operate well in relatively small dataset domains. While Generative Adversarial Networks are a very capable class of models that operate in a way that is very desirable for generative purposes—it just generates output—the fact that the models typically require large amounts of data limit the use cases for many small dataset domains. In the following chapter we will discuss a class of approaches that almost entirely moves away from the input-to-output via learned differentiable functions that we have been discussing for the last three Neural Network focused chapters. Instead, the models learn a policy of how to act in certain situations—**Reinforcement Learning**.

Reinforcement Learning PCG

10

All the way back in Chap. 2 we discussed search-based PCG (SBPCG). This was an approach to generate game content by authoring a space of possible content, a way to move through that space (metaphorically, literally it involves editing some current piece of content), and then some way of evaluating the content (a quantitative measure of content quality or whether it satisfied some set of properties). This allowed a generator to start from some initial (empty, random, etc.) piece of content and make changes to reach some final output content. As with all classical PCG approaches, this approach allows for a great deal of control, but also requires specialized knowledge to apply effectively and a fair amount of authoring effort. Unlike other classical PCG approaches, it also can be very time-consuming due to the iterative changes of the search process. Back in Chap. 4 we gestured at the existence of PCGML approaches that similarly relied on an authored measure of content quality instead of training on existing data. In this chapter we will cover those approaches, which fit under the header of **PCG via Reinforcement Learning** or PCGRL [92].

While the majority of machine learning (ML) approaches covered in this book require existing data, reinforcement learning approaches instead learn from data that is generated via interaction. Instead, these approaches require a number of components. First, PCGRL needs an **environment** to act in. For our purposes, you can think of this like a virtual design studio. It also requires a set of **actions** that an agent can do in said environment. These can be thought of as equivalent to the neighbor functions in a search-based approach, i.e., the particular changes or additions an agent can make to a piece of content. Finally, PCGRL requires a measurement or signal of quality called a **reward** that an agent receives when taking actions. We'll get to the specifics of how this works by building to them piece-by-piece.

The major difference between SBPCG and PCGRL is that, once trained, a PCGRL agent can generate content much more quickly than an SBPCG approach. This is due to the fact that instead of having to search through the space of possible content when we want to

M. Guzdial et al., *Procedural Content Generation via Machine Learning*.
Synthesis Lectures on Games and Computational Intelligence,
https://doi.org/10.1007/978-3-031-16719-5_10

generate something new, PCGRL has learned the best actions to take in a wide range of scenarios, letting it make quick iterative edits that ultimately produce a piece of content that achieves the desired goals. If we imagine we're trying to navigate a city, a search-based approach would have to construct a route from scratch. In comparison, an RL approach already knows many routes through the city and so just needs to follow one.

While this chapter represents an introduction to reinforcement learning, this is only a small subsection of this area. In particular, we focus on the most common RL approaches applied to PCG thus far. Many RL algorithms have not yet been applied to PCG at the time of writing this book, leaving the future application of these approaches to PCG an open problem (we cover other open problems in Chap. 12). We highly recommend seeking a more general RL textbook if this topic is of interest [201], as we lack the space for a complete overview.

10.1   One-Armed Bandits

A common metaphor used to introduce the basic concepts of reinforcement learning is the "one-armed bandit." This metaphor is based on a kind of slot machine with one long arm, often stylized as a bandit. As with all slot machines there is a very, very low probability of success: getting a jackpot. If we somehow knew these odds we could determine roughly how many pulls (i.e., uses of the machine) on average we'd have to do before getting a jackpot. But the manufacturers and operators of these machines would prefer that players not know this probability of success and so keep it hidden.

Now if we played any one machine long enough (wasting a huge amount of money) we could eventually estimate the probability of success. We would simply divide the number of successes by the total number of attempts. As the number of attempts approached infinity, we would approach certainty about the correct probability. That's all well and good, assuming we had infinite money. But what if the situation were slightly different? What if we instead had three machines to choose between? Let's call our three machines A, B, and C, perhaps because they're all located in an alphabet-themed casino. We call this scenario a **multi-armed bandit**, since we now have three arms to worry about!

Ideally, we'd like to maximize our total winnings by only playing the one of these three machines with the best payout. In other words, we only want to play the machine that will lead us to getting the most money, the machine with the highest jackpot and highest probability of success.

Assume for a moment that all three machines have the same jackpot, so this only becomes a question of which of the three has the highest probability of giving us that jackpot. We could use the same idea of just playing each machine infinitely until we find the correct probability, but that introduces a new problem. Every time we use one of the "wrong" machines is a time we could be playing the "right" machine in order to maximize our winnings. So what do we do?

One strategy might be to always try the machine that we think has the highest probability (choosing randomly if they are tied). This will mean at first, when we don't know anything about the machines, we will choose at random. Eventually one of the machines will give us a jackpot, and then we'll just continue to choose that machine forever. Let's call this strategy **exploitation** since we'll be exploiting the first machine that gives us anything. However, this could backfire, as by random happenstance we could get a jackpot on a machine with a worse probability and then end up "trapped" using that machine. The opposite approach, of just randomly **exploring** different machines, also has issues. While we might randomly end up playing the one "correct" machine over and over again, it's much more likely we'll split our attention across this and the two "incorrect" machines. We present a specific example of this in Fig. 10.1. This problem is called the exploration-exploitation dilemma or exploration-exploitation trade-off. We briefly covered a related problem in PCGML back in Chap. 4 in terms of exploration and exploration in generation. These problems are related and both are solved by trying to find some middle ground or combination of the two opposite strategies. Neither strategy is appealing, and the rest of this chapter will introduce modifications to these approaches. Furthermore, we will discuss how even with faulty strategies such as these, we can still learn from them and eventually do better.

Pure Exploitation

Machine Played	Response	Approximated Probability
A	WIN	(A: 1.0, B: -, C: -)
B	LOSS	(A: 1.0, B: 0.0, C: -)
C	LOSS	(A: 1.0, B: 0.0, C: 0.0)
A	LOSS	(A: 0.5, B: 0.0, C: 0.0)
A	LOSS	(A: 0.33, B: 0.0, C: 0.0)
A	LOSS	(A: 0.25, B: 0.0, C: 0.0)

Pure Exploration

Machine Played	Response	Approximated Probability
A	LOSS	(A: 0.0, B: -, C: -)
B	WIN	(A: 0.0, B: 1.0, C: -)
C	LOSS	(A: 0.0, B: 1.0, C: 0.0)
A	LOSS	(A: 0.0, B: 1.0, C: 0.0)
B	LOSS	(A: 0.0, B: 0.5, C: 0.0)
C	LOSS	(A: 0.0, B: 0.5, C: 0.0)

Fig. 10.1 Example of the pure exploitation and pure exploration strategies for the multi-armed bandit problem. We have three machines (A, B, and C) that all have the same payout, but different probabilities of payout (A: 10%, B: 20%, C: 5%). In the pure exploitation strategy (left), if we happen to not get our first jackpot from the best machine, we'll end up stuck playing a worse machine. In the pure exploration strategy (right), even if we get a payout from the best machine, we continue to pick randomly each time

The most common solution to the exploration-exploitation dilemma is to use a strategy that's a combination of both, exploiting sometimes (picking whatever machine we think is best) and exploring sometimes (picking at random). One such approach is ϵ-**greedy**, where ϵ is the probability that we'll explore, exploiting the rest of the time. Thus if $\epsilon = 0.2$ we'll explore 20% of the time and exploit 80% of the time. This allows us to start off exploring since we have no knowledge. Once we get a payout we can then mostly keep playing the machine that gave us the first jackpot. However, thanks to ϵ, if we play for long enough we'll eventually randomly try the machine with a better jackpot probability, at which point we can move over to using that one. There are many other methods for addressing the exploration-exploitation dilemma, but we lack the space to fully discuss them [201]. The important thing is to remember that we have some way to balance between exploring and exploiting in order to ensure the best chance of finding the best machine.

At this point you might be thinking: but what does this have to do with PCGML? Let's consider this metaphor more broadly. The key idea of this jackpot metaphor is that we have some set of choices and unclear payouts or rewards for taking those choices. However, by iterative testing we can eventually settle on what choice will lead to the best possible reward. We can adapt this metaphor to many PCG tasks. For example, what one change to a level would maximize its playability? What one change to a story would maximize its coherency? What one change to a piece of music would maximize its tonality? As long as we can come up with a quantitative measure of quality, we can adapt this metaphor! However, we're still missing something. So far we've assumed we can choose between a single set of three actions and attempt the same actions infinitely. However, this ignores **context**. We introduced context in Chap. 4 in terms of the amount of information we needed to give an ML model for it to make an appropriate decision. It has a similar meaning in this case, where our one-armed bandit example didn't include any context and we'll need to add some to make appropriate decisions in PCGRL use cases, which we'll demonstrate in the below example.

10.2   Pixel Art Example

Let's say we have a simple pixel art RPG and we want to use this method to produce new pixel art creatures. For the purposes of the example, let's go with simple creatures (only black or white pixels) and small images (only 8x8 pixels in size). In this setup each "machine" could represent turning a pixel on or off (when we "use" the machine we'd turn a pixel to white if black or black if white). Imagine we start with the pixel art creature on the left side of Fig. 10.2. We can "play" one of our 64 machines (8x8 pixels), selecting one machine to turn one pixel on or off. We give two examples of this in Fig. 10.2 in the center column, one where we change the pixel at location 0,0 (the pixel in the top left corner) and one where we change the pixel at location 7,0 (the pixel in the top right corner). Both of these middle

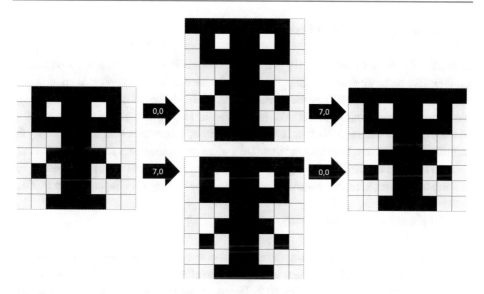

Fig. 10.2 Example of different actions impacting a small pixel character

characters look a bit odd. However, if we make one more change we get a reasonable pixel art character that looks like it has horns or ears. But there's something off here.

When we were using the metaphor of the one-armed bandits, pulling the lever of one machine didn't impact any of the other machines. However now, "pulling a lever"/doing some action changes the context. This means that different actions are dependent on one another, and that we care about sequences of actions, not just individual actions on their own. However, at any point there still might be just one best action to do (one best lever to pull or one best pixel to change). Like in the situation of the middle columns in Fig. 10.2, the best thing to do might be to "complete" the character by making the symmetrical change. To figure out this best action we can use the same approach as we discussed before to find the best machine. By iteratively trying all actions/making all pixel edits we possibly can, we can eventually figure out the best action to take. Therefore, we can think about solving a bandit problem for every possible pixel character: for every possible set of black and white pixels. This is what we meant by context earlier. We call each individual pixel character or each individual set of black and white pixels a **state**. The different machines we can play, the different ways we can impact the state, are called **actions**, and the quantitative measurement of quality is called the **reward**. These are the basic building blocks for a particular problem formulation called a **Markov Decision Process** or **MDP**.

10.3   Markov Decision Process (MDP)

A Markov Decision Process (MDP) is a way of formulating a sequential decision making problem like those we described above (choice of what pixels to turn on/off to create a final pixel character over time). They are named after Andrey Markov, who invented the Markov chain model we discussed in Chap. 6. MDPs can in fact be thought of as extensions of Markov chains where instead of just random transitions, we can take actions that modify the transition probabilities. However, while MDP's often have randomness, it is possible to use them to model deterministic settings (where there's no random chance of the same action leading to different effects), which is the case we focus on in this chapter and in PCGRL more generally.

MDPs have the following primary components:

- S: A set of states called the state space. We use a lower-case s to represent an individual state. For example, the state space is equivalent to all possible 8x8 black and white pixel images from the example in Fig. 10.2. However in other domains it would represent every possible version of whatever content we wanted to generate (e.g., levels, game rules, music, etc.). It is essentially identical to the concept of a search space from search-based PCG, meaning that typically the developer who sets up the MDP defines the state space.
- A: A set of actions called the action space. With the pixel characters this represents what changes can be made to the state or, put another way, how we can move between states. This is equivalent to the concept of a neighbor function for search-based PCG, or a mutation function for evolutionary search specifically.
  Sometimes not all actions will be available to us. For example, if we instead had 3 colors for our pixel images (let's say black, white, and red) then we might say A is 3 actions for each pixel (turn black, turn white, or turn red), for a total of 192 actions (3*64, since there are 64 pixels in our 8x8 image). However, if we were in a state where all 64 of our pixels were red then we wouldn't want to have the "turn red" action available. In this case we represent A as A_s or the set of actions possible in state s. This is similar to the choice of representations described in Chap. 4, in terms of helping to cut down on complexity.
- R: A reward function $R(s, a)$ that returns how much reward or punishment (positive or negative value) an agent gets when it takes a certain action a in a certain state s. You will also sometimes see this written as $R(s, a, s')$ or $R_a(s, s')$ for non-deterministic cases, where we want to model the reward we get when we take action a in a state s and end up in another state s'. For example, if we're in an environment where we're trying to design a pixel monster, but sometimes choosing the "turn red" action makes the pixel turn white or black instead. But in PCGRL we typically don't introduce this kind of randomness, and so we always end up in the same next state when we take the same action, or more technically $R(s, a) = R(s, a, s')$.

The primary challenge of any reinforcement learning problem is maximizing the agent's expected reward over time. Depending on how we set up the MDP we might get a reward only for a "complete" piece of content. For example, if we had an agent writing a story or a piece of music, we might only evaluate the piece of content when it was complete, giving the agent a high value if it's good and a low value if it's bad. Alternatively, we might give the agent feedback after every action it takes. This might make sense in some cases, like if the agent is editing a level from a random starting point, where every single state represents a potential level, some better than others. It is generally easiest to learn in this latter case with more consistent feedback, but this may not be appropriate for all problems or types of content.

- T: While it isn't often used in PCG, a typical MDP also has a transition function T, represented as $T(s, a, s')$ or $T_a(s, s')$, or the probability of transitioning from one state s to another state s' when taking action a. Some denote T as P, but we generally avoid this as P is overloaded in probabilistic approaches.

  T is most helpful when employing an MDP to act in environments like automatically playing a game, or guiding the behavior of a robot or autonomous vehicle. In these situations taking an action doesn't always lead to the expected next state and so it's important to model this uncertainty. For example, if we were using an MDP to model the behaviour of a robot, we might expect picking the action to go forward to lead us to going some distance forward. However, if our robot's feet or treads get caught on something we might end up stuck in the same state. We could model this as saying there was a 90% chance of going forward and a 10% chance of staying in the same spot ($s=s'$). However, this has not been traditionally employed for PCG problems where we assume that actions are discrete and deterministic (each action does what we expect every time).

S, A, R, and T are the traditional parts of an MDP. An MDP is defined with one goal in mind: learning a **policy**. A policy, represented with the symbol π, is a mapping of states to actions $\pi(s) \rightarrow a$ that tells us which action to take in a given state. A policy doesn't necessarily mean anything on its own. For example, we could have a policy that says for every state we should do nothing or one where we always take the same action. Because of this, it's the typical goal of an MDP to learn π_{max}, the policy that maximizes the expected reward. Processes for learning this π_{max} are collectively referred to as **Reinforcement Learning**.

There are many RL algorithms for learning an optimal policy given a Markov Decision Processes (MDP). However, they all follow the same general outline given in Fig. 10.3. An RL agent iteratively takes actions (a_t) in an environment defined by an MDP. The environment uses the internal transition function and reward functions to return the next state (s_{t+1}) and reward (r_{t+1}) respectively. The RL agent uses the reward signal (r_{t+1}) to iteratively improve its policy $\pi^t \rightarrow \pi^{t+1}$ attempting to find the actions that will maximize its reward. This process then continues, with the agent taking another action, receiving another state and reward pair and so on. In Fig. 10.3 we give an example of a specific action and the state

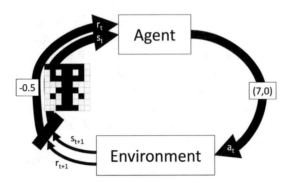

Fig. 10.3 Standard RL setup with an agent receiving a new state along with a reward for some (unseen) prior action from the environment, prompting the agent to take a new action, and then this loop continuing into a new state and reward (s_{t+1} and a_{t+1}). Each arrow includes one example from Fig. 10.2 for a reward, state, and action

we might get in return, taken from the example in Fig. 10.2, along with an arbitrary reward value the agent might have received (-0.5).

There are many ways to try to go about learning the optimal policy. One common way, which we'll investigate further below, is to try to determine the **value** of each action for each state. This value can be thought of as the cumulative reward we'll get if we keep following the current policy (doing whatever action the current policy tells us to do in every future state). We can represent this value of each action in each state as a dictionary or look-up table called a **Q-table**. This data structure is typically represented as $Q(s, a)$, and can be thought of as a table indexed on states and actions, where each cell represents the expected cumulative value of taking a particular action in a particular state and then continuing to follow the policy. If we have an accurate Q-table, one that accurately represents the value of each action, then we can easily extract the optimal policy by finding the action with the highest value for every state: $\pi_{max}(s) = \text{argmax}_a Q(s, a)$. The challenge then is learning an accurate Q-table, with a common approach being the aptly named **Q-learning**. We'll discuss this further below, but first we'll introduce a specific MDP example to help clarify how to define an MDP for a particular problem.

10.4   MDP Example

Now that we've introduced the components of an MDP and the goal of reinforcement learning in terms of achieving π_{max}, let's look at an example of one application of reinforcement learning to PCG. Let's say that we're generating creatures for a turn-based monster battling game like Pokémon, Digimon, or Yokai Watch. To keep things simple, imagine that in this case we're only interested in their characteristics or statistics, and assume we're leaving the visuals and sound effects to another process (perhaps something like the earlier Figs. 10.2 and 10.3 for example). We can then define each monster with the following characteristics:

1. **Health**: The health of the monster, if it reaches 0 then they lose the battle. Initial health has a minimum value of 1 and a maximum value of 100 to start.
2. **Armor**: An extra layer of protection for the monster. When it would take damage due to an attack, it instead takes damage$' = \max(0, \text{damage} - \text{armor})$ damage to health. Meaning that if the damage<armor then the attack is nullified completely. Armor can range from 0 to 50.
3. **Speed**: The ability of the monster to dodge. This characteristic is equivalent to the probability that the monster dodges an incoming attack. So if the speed value is 50 then there's a 50% chance that the monster dodges an incoming attack. Let's say it also determines the order of attacks in that the monster with the higher speed value attacks first (or randomly in the case of a tie). Speed can range from 0 to 50.
4. **Damage**: The damage value of each monster. This represents the amount of damage that a monster would do to another monster's health if the target monster doesn't dodge or have armor. Damage can range from 1 to 50.

These four statistics would then define all possible monsters that could possibly exist in our game. While this is much simpler than a real monster battling game, this is still a lot of possible monsters! Specifically it is:

$$|\text{Health}| \times |\text{Armor}| \times |\text{Speed}| \times |\text{Damage}| =$$
$$100 \times 51 \times 51 \times 50 = 13,005,000 \text{ monsters}$$

This represents our **State space (S)**. But not all of these monsters will be perceptually distinct. Consider this, if two monsters are the same but one has a health of 41 and the other has a health of 42, is that really all that different? But even if we assume that monsters need to at least have a difference of 5 for some characteristic to seem meaningfully different that's still 20,000 perceptually distinct monsters ($20 \times 10 \times 10 \times 10$).

We give an example of four such possible monsters in Fig. 10.4. We note again that we're focusing on the characteristics (underneath the pixel images) only in this example. We include the pixel images to help convey what these values suggest about the monsters. In the example we have an "average" monster (with all stats at 50% of the max) at the top left, a "tank" monster focused on defense (Heath: 100, Armor: 50, Speed: 5, Damage: 10) on the top right, a "weak" monster on the bottom left with all the lowest possible values, and finally we have a fast "glass cannon" monster at the bottom right (Health: 20, Armor: 0, Speed: 50, Damage: 50). Each one of these represents one state s in our state space S.

Now that we have our state space S, we next need to define our action space A and reward function R. In this case, as with most PCGRL examples, we'll ignore the transition function

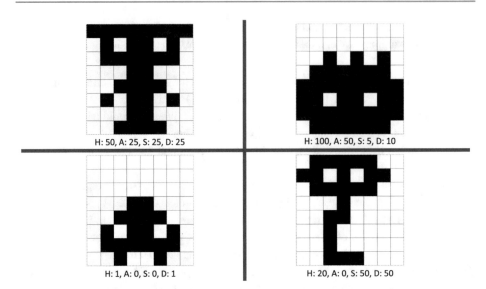

H: 50, A: 25, S: 25, D: 25

H: 100, A: 50, S: 5, D: 10

H: 1, A: 0, S: 0, D: 1

H: 20, A: 0, S: 50, D: 50

Fig. 10.4 Four example monsters that represent states in our example MDP (note the generation of the pixel visualizations is assumed to be a separate MDP). Top left is our "average" monster, top right is our "tank," bottom left is our "weak" monster, and bottom right is our "glass cannon"

T and assume our transitions are deterministic. For the action space A we can simply define it as adding or subtracting one to each characteristic (e.g., adding or subtracting 1 to health, armor, speed or damage) unless we're at the minimum or maximum allowed value.

The reward function R is much trickier for this problem. As with a fitness function or heuristic for search-based PCG, a single function to convert a piece of content to a numerical value of quality is not always possible. Indeed while it might seem obvious that the "average," "tank," or "glass cannon" monsters in Fig. 10.4 are more valuable than the "weak" monster, this depends on context. For example, what if we wanted to generate easy monsters that show up at the beginning of the game? Or if our game has a notion of growing or evolving, what if we wanted a baby version of a later monster? That said, if we're looking to generate boss monsters this "weak" monster would certainly be less appropriate. For now, we'll sidestep this problem and assume we're focused on the problem of generating balanced monsters. By balanced here we mean that in a fight against the "average" monster from Fig. 10.4 (Health: 50, Armor: 25, Speed: 25, Damage: 25) a "balanced" monster should win roughly 50% of the time.

Specifically for our problem let's say that our reward function R takes in a state $R(s)$ representing a monster. This is notably distinct from our earlier reward function definition $R(s, a)$, but this is due to the fact that we don't care what action we took to get into a state– to find a particular monster. Our reward function can then simulate one million games against the "average" monster, which can be done very quickly since there's no decisions to be made (every monster just attacks every turn in our simple game). From there we can compare the win rate we achieve to our goal 50% win rate and get a single, numerical value out of this process as our reward function R. For now we can assume a simple formula for our reward function derived from this win rate:

$$r = \frac{50 - |winrate - 50|}{50} \tag{10.1}$$

With this reward function the maximum value is 1.0 when we have a perfect 50% win rate and 0.0 for either a 0% or 100% win rate. With this, we have everything we need to set up our Q-learning application.

10.5   Tabular Q-Learning

As a reminder, the focus of reinforcement learning approaches is to identify π_{max}, the optimal policy. Q-learning is so named because it focuses on iteratively building up (learning) the Q-table we introduced earlier. A Q-table, typically represented as $Q(s, a)$, represents the expected value of taking action a at state s. Since we have 8 actions (each characteristic going up and down) at each state, this means we have a table of size 13,005,000 x 8 (state x action). We start by initializing this table, for example with all zeroes, all ones, or some type of random initialization. A random initialization is generally preferred as it's more likely to be closer to the true Q-table we're hoping to learn. However, to make things simpler, we'll go with all zeroes. Remember that our goal is to find the Q-table where the values of each action for each state accurately represent the cumulative reward the agent would receive if it continued to take actions according to the Q-table (the action with the highest value in each state). Once we find this Q-table we can easily identify the optimal policy. To attempt to learn this ideal Q-table, we need to start collecting training information.

Recall the multi-armed bandits example from the beginning of the chapter. The goal there was to find the probability of payout for different machines. The metaphor here is that we're trying to find the "payout" (expected reward r) of using a particular machine (action a) in a given context (state s). To learn this, we need to start taking actions in the state space. In other words, we need to start designing monsters!

When applying Q-learning or other RL algorithms, we typically start from some initial state (a random monster in our case) and take a sequence of actions. This sequence of actions is called a **rollout**. A rollout typically ends when either it reaches a terminal state or when it reaches some developer-defined maximum number of actions (called the rollout length, sometimes represented with l). Our environment doesn't have terminal states, but if we were producing a piece of music or a story with a pre-defined length or where the RL agent could choose to add an endpoint, then this might be relevant.

10.5.1 Rollout Example

Let's run a single rollout as an example. To keep things reasonable let's say we only have a rollout length l of 6, though this is rather small. Let's say our initial state is the "glass cannon" monster from Fig. 10.4 with health $= 20$, armor $= 0$, speed $= 50$, and damage $= 50$ (H : 20, A : 0, S : 50, D : 50). When picking the actions to take during the initial rollout we don't know anything, so we can take them at random. This initial state has a winrate of 36.81% against the average monster. That means our initial reward is 0.736 out of a possible 1.0, which is pretty good! But we can do better. Let's say we took the following rollout steps from this initial state.

1. **Health +1**: (H : 21, A : 0, S : 50, D : 50); winrate $= 36.81\%$; reward $= 0.736$
2. **Health +1**: (H : 22, A : 0, S : 50, D : 50); winrate $= 36.72\%$; reward $= 0.735$
3. **Health +1**: (H : 23, A : 0, S : 50, D : 50); winrate $= 36.79\%$; reward $= 0.736$
4. **Health +1**: (H : 24, A : 0, S : 50, D : 50); winrate $= 36.75\%$; reward $= 0.735$
5. **Health +1**: (H : 25, A : 0, S : 50, D : 50); winrate $= 36.77\%$; reward $= 0.735$
6. **Health +1**: (H : 26, A : 0, S : 50, D : 50); winrate $= 84.00\%$; reward $= 0.399$

In this case we're pretending that the RL agent randomly "decided" to just pick the health+1 action six times in a row. As we can clearly tell the winrate (and therefore reward) hovers around the same number, but the sixth step leads to a drastic change. The first five steps still change by a bit because of small variations in the simulation of battles, but those aren't significant. If you think about this for a second it makes sense, since the "average" monster has a damage of 25 it could previously take out the "glass cannon" in one shot. The "glass cannon" was only doing reasonably well because it had a 50% chance to dodge. If it failed to hit the "average" monster once (fairly likely with the "average" monster's 25% change to dodge), it would be taken out in one shot if it didn't dodge. However, as soon as we bring the glass cannon's health up to 26 suddenly it can take an extra hit before being taken out.

At this point what we need to do is to use this rollout to update our Q-table. We do this with what's called the **Q-update function**. We'll cover this in the next section. As a warning, this function is the most technically dense part of this chapter, but we'll break it down with an example afterwards.

10.5.2 Q-Update

The Q-update function is as follows:

$$Q^{new}(s_t, a_t) \leftarrow Q(s_t, a_t) + \alpha * (r_t + \gamma * max_a Q(s_{t+1}, a) - Q(s_t, a_t)) \qquad (10.2)$$

$Q^{new}(s_t, a_t) =$ The new Q-table value for s_t and a_t

$Q(s_t, a_t) =$ The old Q-table value for s_t and a_t

$\alpha =$ Distinct from a, α is the learning rate

$r_t =$ The reward we just received for taking a_t in s_t

$s_{t+1} =$ The new state

$\gamma =$ The discount factor, the weight given to the values of s_{t+1}

This equation is a bit dense, and it introduces a couple of new variables, so let's walk through it. $Q^{new}(s_t, a_t)$ represents the new value in our Q table for this particular state s_t and action a_t after the Q update function. $Q(s_t, a_t)$ represents the old value in our Q table for this particular state s_t and action a_t. Since we specified that we were initializing our Q-table with 0s this means that this is just equal to 0 at first. α is a new variable and it represents the **learning rate** (sometimes represented with η), which is the rate at which we'll be changing the Q-table. This serves the same purpose as the learning rate introduced for DNNs in Chap. 7. This number is typically low, in order to avoid the danger of a randomly bad or randomly good run impacting things too much. For now, we will set α to 0.1, but the optimal value of this hyperparameter depends on the domain.

We then have an expression in parentheses, which is referred to as the **temporal difference** representing the change in our Q-table. First up, we have r_t, which is the reward we just got when we took action a_t in state s_t and wound up in state s_{t+1}. This is equivalent to the reward that we received when we got to this state at time step t of the rollout. From there we have another symbol γ, referred to as the **discount factor**, which basically indicates how much we pay attention to the future.

If γ is low then we don't pay much attention to the future, and if it's 1.0 then we treat the future as equivalent to the present. We're multiplying γ times $max_a Q(s_{t+1}, a)$ or the value of the maximum valued action we can take in the next state s_{t+1} we just got to according to our Q-table. So when γ is 0.0 then we ignore this value.

Let's say we used a γ of 0.5, but again this is a hyperparameter where the optimal value is dependent on the particular domain. Notably, since we need to know what happens in the "future" we'll be updating our Q-table from the end of the rollout to the beginning of the rollout, as we'll demonstrate in a moment. Finally, we subtract this value by $Q(s_t, a_t)$ because we're interested in representing the change in the current Q-table value for this state s_t and action a_t. Given this we can now demonstrate using the Q-update function after our earlier rollout. So, putting it all together, the Q-update function updates the value of the current action-state pair by looking at a combination of the current reward, the possible max value of the next state, and the value of the current state.

10.5.3 Q-Update Example

As we mentioned earlier we run our Q-update function once per rollout step from the end of the rollout to the beginning. This can be a bit confusing so we visualize it in Fig. 10.5. All six rollout states happen first, and then all six Q-update steps happen, going "backwards" along the rollout.

1. We start with the second-to-last rollout state, since we need a next state as part of the Q-update function. As a reminder we initialized our Q-table with 0s so we can go ahead and replace every $Q(s_t, a_t)$ with 0. Therefore, this update function step looks like $Q^{new}(s_t, a_t) \leftarrow 0 + \alpha * (r_t + \gamma * 0 - 0)$ or just $Q^{new}(s_t, a_t) \leftarrow \alpha * (r_t)$.
   s_t in this case is (H : 25, A : 0, S : 50, D : 50) (the second-to-last state) a_t is Health+1 (the last action), and r_t is 0.399. Since we decided α is 0.1 the new value of $Q(s_t, a_t)$ or $Q((H : 25, A : 0, S : 50, D : 50)$, Health +1) is 0.04. This might not seem like a lot but it's better than 0!
2. From there we consider the third-to-last rollout state (H : 24, A : 0, S : 50, D : 50). While once again we can replace all instances of $Q(s_t, a_t)$ with 0, we now have a value for $max_a Q(s_{t+1})$ because we just calculated it (0.04). This might be a bit confusing, but since we went from a state with H : 24 to a state with H : 25 due to taking action Health+1, the state with H: 25 is the s_{t+1} or next state from the state with H : 24 (s_t). Given all of this our Q-update 2 is $Q^{new}(s_t, a_t) \leftarrow 0 + \alpha * (r_t + \gamma * 0.04 - 0)$ or since $\alpha = 0.1$ and $\gamma = 0.5$ then Q-update 2 is $Q^{new}(s_t, a_t) \leftarrow 0.1 * (r_t + 0.5 * 0.04)$. Since s_t in this case is the (H : 24, A : 0, S : 50, D : 50) state, a_t is Health +1, and r_t is 0.735, then the new value of $Q((H : 24, A : 0, S : 50, D : 50)$, Health +1) is $0.1 * (0.735 + 0.5 * 0.04)$ or 0.076.

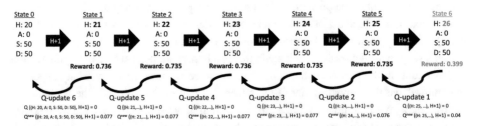

Fig. 10.5 The example of the rollout and the subsequent Q-update function calls given below. The red text at the end is only to bring the reader's attention to specific values

3. We can then use the same approach as in the previous step for this and all future Q-updates, with the only difference being the $max_a Q(s_{t+1})$ and small changes in r_t. So for Q-update 3 we get $Q((\text{H}:23,\text{A}:0,\text{S}:50,\text{D}:50), \text{Health}+1)$ is $0.1 * (0.735 + 0.5 * 0.076)$ or 0.077.

4. In Q-update 4 for the Q-table cell $Q((\text{H}:22,\text{A}:0,\text{S}:50,\text{D}:50)\ \text{Health}+1)$ gives us $0.1 * (0.736 + 0.5 * 0.077)$ or 0.077.

5. In Q-update 5 for the Q-table cell $Q((\text{H}:21,\text{A}:0,\text{S}:50,\text{D}:50), \text{Health}+1)$ gives us $0.1 * (0.735 + 0.5 * 0.077)$ or 0.077.

6. Finally for $Q((\text{H}:20,\text{A}:0,\text{S}:50,\text{D}:50), \text{Health}+1)$ we get $0.1 * (0.736 + 0.5 * 0.077)$ or 0.077.

This might seem like a lot of work for nothing, but by iteratively running rollouts we'll eventually converge on a Q-table where if we take the action a that maximizes $Q(s, a)$ for some state s then we'll end up taking the optimal action. Optimal action in this case meaning the action that will change our current monster to be closer to a balanced one. Let's show one more rollout as an example, though we typically require several orders of magnitude more rollouts to achieve reasonable results.

10.5.4 Rollout Example 2

Let's say we do another rollout, meaning another six random steps from our initial state $(\text{H}:20,\text{A}:0,\text{S}:50,\text{D}:50)$ (though notably in this environment we'd likely reset our initial state to ensure we have better coverage, especially with such a low rollout length l). Let's say we're using the ϵ-greedy function from earlier (the exact value of ϵ doesn't matter) so we're more likely to take the action a in each state with the highest value. Since almost all values equal 0, this means we'll likely repeat the same steps as in our first rollout example. Let's say we repeat the same steps as before except for the last one.

1. **Health +1**: $(\text{H}:21,\text{A}:0,\text{S}:50,\text{D}:50)$; winrate $= 36.81\%$; reward $= 0.736$
2. **Health +1**: $(\text{H}:22,\text{A}:0,\text{S}:50,\text{D}:50)$; winrate $= 36.72\%$; reward $= 0.735$
3. **Health +1**: $(\text{H}:23,\text{A}:0,\text{S}:50,\text{D}:50)$; winrate $= 36.79\%$; reward $= 0.736$
4. **Health +1**: $(\text{H}:24,\text{A}:0,\text{S}:50,\text{D}:50)$; winrate $= 36.75\%$; reward $= 0.735$
5. **Health +1**: $(\text{H}:25,\text{A}:0,\text{S}:50,\text{D}:50)$; winrate $= 36.77\%$; reward $= 0.735$
6. **Speed −1**: $(\text{H}:25,\text{A}:1,\text{S}:49,\text{D}:50)$; winrate $= 35.80\%$; reward $= 0.716$

The final action we took led to a final state with a slightly improved reward relative to the first rollout example. We won't run the full Q-update step this time but because the reward we got for taking action Speed -1 is better than for taking action Health +1 in state $(\text{H}:25,\text{A}:0,\text{S}:50,\text{D}:50)$ we'll be more likely to do the former again in the future. This is one tiny positive change towards more optimal behavior, but after enough of these small positive changes we can arrive at a Q-table where the maximum action for each state is the optimal one.

10.5.5 Tabular Q-learning Wrap-up

By iteratively running rollouts (using ϵ-greedy and our current Q-table) and then updating the Q-table we can eventually (after potentially millions of steps) end up with a Q-table that reasonably approximates the true value of each action. We can then repeatedly query the Q-table to go from a random state (monster) to a more balanced state (a different monster). For example, the glass cannon monster would eventually converge to $(H : 40, A : 4, S : 47, D : 45)$ with a winrate of roughly 50%. This process of querying the Q-table is very fast, to the point where we could use it at runtime in a game even with very little computation power available. The final "balanced" monsters are also non-obvious due to the (relative) complexity of the problem. However, the process of applying Q-learning to this domain would take far too many iterations to make it worthwhile, and a human designer could just as easily keep querying the reward function until they found a sufficiently balanced monster. That brings us to the question of how we could speed up this process? There are other tabular reinforcement learning techniques, such as Value Iteration [68, 201] (learn the value of states instead of actions) and Policy Iteration [16, 201] (ignore values and just iterate on policies directly), but none of them directly solve this problem.

10.6   Deep Q-Learning

In 2013, Mnih et al. published "Playing Atari with Deep Reinforcement Learning" [127], a foundational paper with a simple pitch: Q is a function (that provides the expected value of following our policy in a given state), what if we learned that function directly (using a universal function approximator—a NN as in Chaps. 7, 8, and 9) instead of using a table? The intuition is that many states are functionally the same, meaning that for tabular reinforcement learning we end up tracking multiple states that are basically identical, which is a waste of computation. This is true for the example in the last section, consider the states we ended up in for rollout steps 1-5, with essentially identical winrates and rewards. With tabular Q-learning, nothing we learn from $(H : 23, A : 0, S : 50, D : 50)$ carries over into $(H : 24, A : 0, S : 50, D : 50)$ despite the two states being mechanically identical. The size of the Q-table is also an issue, with potentially millions of points we have to store in memory, even for simple MDPs like the one above.

Instead of a literal table, a DNN can allow us to learn what states can be treated as the same, and which states stand out. DNNs are, after all, just function approximators. Therefore they can learn to approximate the Q table "function" (given an a and s return a value). This drastically sped up the training process of Q-learning in many domains, such as the Atari game playing domain from the original paper. Hearing that this is faster than tabular Q-learning may be surprising if you're familiar with issues like how high-profile applications of deep RL have trained for amounts of time longer than the human civilization has existed (by training in parallel) [15]. However, this just demonstrates how difficult these problems are, and how utterly impossible they would be to address with tabular reinforcement learning.

The setup for applying deep reinforcement learning is still very much like applying tabular reinforcement learning. We still need a representation of our state space S, actions A, and reward function R. However, we also need to design the DNN that will act as the RL agent (a Q-table for Q-learning, a policy network for policy iteration, etc.). The design of these networks is typically fairly specialized to the problem domain, often combining elements of neural networks for image processing (as discussed in Chap. 9) and sequence processing (as discussed in Chap. 8). Designing good networks is an open research problem , because of this the most common approach is to adapt a DNN from prior work to your problem or to reuse it directly.

Deep Q-learning shares the same premise as Tabular Q-learning, but with key differences in training. First, we gather results from a large number of rollouts called a **training episode**, instead of single rollouts. In Eq. 10.2 we saw how a Q-table would be updated, and it is similar for a Deep Q-Network (DQN); However, instead of updating the Q-values, we are training a function that tries to predict the Q-values. At its core, this is just another regression, except instead of trying to predict known labels (as in Chap. 7), we are trying to iteratively improve the estimated Q-values. First, we predict the estimated Q-value ($\hat{Q}(s_t, a_t)$) based on the observed reward and the expected future reward . This is then the target value in a regression (with Euclidean loss) with the prediction being the current Q-value ($Q(s_t, a_t)$):

$$\hat{Q}(s_t, a_t) = r_t + \gamma * max_a Q(s_{t+1}, a) \tag{10.3}$$

$$L = (\hat{Q}(s_t, a_t) - Q(s_t, a_t))^2 \tag{10.4}$$

This has many similarities to the tabular Q update, except (1) the use of α to modify the update has been replaced by a NN training step, and (2) Q is a neural network. We then repeat this process until our DNN converges, just as we would for a supervised learning problem. Unlike a supervised learning problem, our training set will change as our agent learns, altering how it interacts in the environment. Eventually, after enough training episodes our DNN will converge (in most cases). As we noted in the Q-update example in Sect. 10.5.3, it is common for it to take a number of Q-updates for reward to filter back to an early state. But this filtering back is extremely important, as the policy should reflect that actions at an early timestep might eventually lead to a high reward later on. To achieve this with Deep Reinforcement Learning, it is very common to keep a **replay buffer**. This replay buffer is simply a list of all actions, states, and rewards that have been observed during all episodes. When training, instead of just using the most recent data, the replay buffer is sampled and all of the sampled values are used to update the policy. With R as the replay buffer, the loss is then calculated as such:

$$R = \begin{bmatrix} s_0, a_0, r_0 \\ s_1, a_1, r_1 \\ ... \\ s_t, a_t, r_t \end{bmatrix} \tag{10.5}$$

$$L = \sum_{s_i, a_i, r_i \in sample(R)} (r_i + \gamma * max_a Q(s_{i+1}, a_i) - Q(s_i, a_i))^2 \tag{10.6}$$

Using a replay buffer means that earlier portions of the episodes will be revisited, allowing them to be updated, improving the training of the policy. Beyond simply uniformly sampling from the replay buffer, it is also possible to weight the sampling to prioritize different states more or less than others. This is beyond the scope of this book but interested readers can find work that details how this prioritization might occur [153].

All of the examples in the section have come from the viewpoint of Deep Q-learning, but just as in tabular Reinforcement Learning, Q-learning is only one possible approach. Another common approach for Deep Reinforcement Learning is **Actor-Critic** learning. In Actor-Critic learning, there are two networks: the Actor that predicts which action to take, and the Critic that predicts the value of the Actor's policy. This is different from DQNs which learn one network that predicts the Q-value of the policy and then uses that Q-value as the basis for the policy. The Critic is updated much as the DQN is, it iteratively tries to update the predicted value of the policy. The Actor uses the gradient of the Critic to improve its performance (updating so as to maximize the Critic's estimated value). This setup can be useful in domains where the actions come from a continuous range as opposed to discrete values. For instance, instead of choosing a stat to update by ± 1, the Actor could choose any arbitrary value for each stat, allowing it to move through the state space S much faster. Details into Actor-Critic methods are beyond the scope of this book, but the interested reader can find resources elsewhere [126, 222].

The other benefit of Deep RL is generalization. Because our DNN should learn to handle multiple states in the same way, this is equivalent to generalizing over these states. This means that we'll be more likely to respond appropriately to similar states that we didn't see during training. However, this doesn't entirely solve the generalization problem, if we somehow missed a whole section of the state space (which can happen!) we'll still be unlikely to behave optimally. For example, imagine if, by random happenstance, we happened to entirely miss the area of the state space around the "tank" monster in Fig. 10.4. At that point, if we saw the "tank" monster during testing, our DNN would be unlikely to take appropriate actions to balance it.

10.7   Application Examples

This chapter thus far has been fairly abstract. We've tried to keep the running example as a way to keep things grounded, but it can be hard to imagine how or when to apply PCGRL. Thus, in this section we give a small set of examples of ways PCGRL has already been employed by PCGML researchers. We lack the space to cover these approaches in detail, but you should now have the necessary technical background to go read the cited papers for that detail.

There are basically two strategies when it comes to applying RL to PCG problems, centered on the question of how to handle a reward function. The first of these essentially treats PCGRL like search-based PCG (SBPCG), attempting to define a reward function that evaluates content quality like the heuristic of a SBPCG problem. For example, there's the work of Khalifa et al., who originally coined the term "Procedural Content Generation via Reinforcement Learning (PCGRL)" [92]. In this paper the authors attempted to generate levels for three domains (Binary, a one-room simplified version of Zelda, and Sokoban) and employed a measure of playability along with some rules as the basis of their reward function. This was followed up by many of the same authors in a work that extended this approach to two additional domains with a reward function that instead measured the closeness of generated levels to existing levels given a set of target metrics [40]. Similar to SBPCG, this work requires a well-authored and domain-dependent reward function. Even when comparing to existing content, the developer of the system still has to define what metrics to consider in the comparison (size, patterns, solution lengths, etc.) However, unlike SBPCG once it is trained a PCGRL agent can generate content very quickly, while a SBPCG approach can take a long time to output a particular piece of content.

The second approach for applying PCGRL is to instead use human feedback as (part of) a reward function. This can be thought of as analogous to some of the few deployed SBPCG systems, such as those used in Petalz [145] and Darwin's Demons [186]. These systems chose to make use of human feedback as part of their fitness function or heuristic, either due to viewing a human-authored fitness function as difficult or impossible (how do you quantify fun?) or due to a desire to adapt to particular individuals. One of the earliest approaches to attempt to employ human feedback to optimize an MDP was the work of Thue and Bulitko, who optimized an MDP to adapt an interactive narrative to a player [206]. Essentially, if a player liked a particular kind of narrative, the system would attempt to deliver more of that kind of narrative. However, there's some debate as to whether this "counts" as reinforcement learning, given that the author's model the player's policy but do not attempt to learn one of their own.

Similarly, Guzdial et al. employed a pre-trained model and RL to adapt to designers [57]. A screenshot of their tool "Morai Maker" can be seen in Fig. 10.6. The tool allowed a human designer and an RL agent to interact, taking turns making additions to a Mario-like level. The RL agent's reward was based on if its additions were kept (positive reward) or deleted (negative reward). Professional developers made use of the tool in a human subject study and found that the tool adapted to their particular design style. Technically, what was happening was that the RL agent was able to learn a unique policy that better fit its design partner in comparison to the initial policy. A system like this could, in theory, be used to adapt a generator to a player's preferences, similar to the approach of Thue and Bulitko [206].

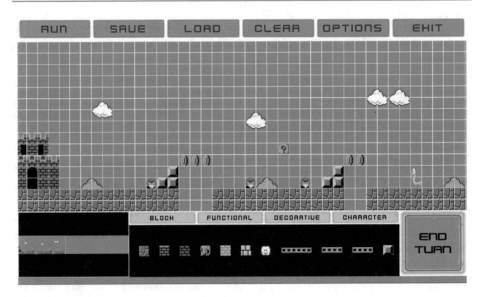

Fig. 10.6 Example of the editor for "Morai Maker"

10.8   Takeaways

In this chapter we introduced PCGRL and dug into how to address a PCG problem with RL. Reinforcement Learning has had amazing success in the automated playing of games, but has yet to replicate that success in the automated designing of games. While there have been some prototype tools for level design [35, 57], this has been the extent of PCGRL's application to games thus far. However, with new PCGRL research systems being developed regularly, there's a sense that we've only scratched the surface of what is possible with this technique [119, 130].

Unlike other PCGML approaches, generating novel content with PCGRL is fairly straightforward, since all you need to do is define your MDP. However, the question of a reward function looms large. Like SBPCG before it, PCGRL has seen a great deal of academic interest, but it's proven difficult to quantify what high quality content looks like in a reward function. Given the decades that have been spent trying to quantify or define what makes a game "fun" it's unlikely that we'll solve this problem in the near future. However, with approaches that compare generated content to existing content [40] or employ human feedback [57, 119] there may be future developments that allow us to solve, or at least largely sidestep, this problem.

PCGRL isn't alone in potentially relying on human feedback, of course. Many researchers see the future of PCGML in smart tools rather than purely autonomous content creators. These systems would help designers during the development of a game, rather than attempting to produce content on their own. In this way we can attempt to marry the positive qualities of ML (pattern recognition, fast iterations, etc.) and human designers (evaluating quality, having a sense of context, etc.). In the next chapter we'll dig into these kinds of mixed-initiative systems.

Mixed-Initiative PCGML

11

In the previous chapters we introduced different machine learning paradigms and discussed how they could be leveraged for PCGML. What each of those previous chapters had in common was that they treated training and generation as static processes. That is, model training and content generation were processes that someone would start and then wait for the outcome. In this chapter we will introduce a different viewpoint that positions user input and interaction more prominently during training and generation. This viewpoint is helpful when thinking about designer and user-facing tools that leverage PCGML techniques. Specifically, this chapter introduces **Mixed-initiative PCGML** (sometimes **co-creative** instead of mixed-initiative). Lai et al. [103] describe mixed-initiative content creation methods as those methods that allow for agents (human or AI) to take turns contributing to the creation of a piece of content. Therefore, mixed-initiative PCGML describes PCGML systems where the user is interacting with, and potentially collaborating with, the underlying PCGML models in order to refine the model or control the generation. The idea behind mixed-initiative PCGML is a human designer will almost always have more context, domain knowledge, and insight into what is needed in a design than a stand-alone PCGML model could, and so we should leverage that knowledge and insight.

In some sense the PCGML approaches we discussed up to this point can also be considered mixed-initiative, where the user interaction is placed at the very start of the pipeline (i.e., choosing the training data and model setup) and at the very end of the pipeline (i.e., choosing the desired content from the generated pieces). However, in this chapter we will delve into potential systems where the user interaction is situated more centrally in the pipeline. It is important to note that mixed-initiative and co-creative PCGML is a greatly underexplored area of research, and as such a lot of the topics we discuss in this section are closer to best practices at the moment, and should not be taken as definite design rules for building out these systems. This chapter is more meant as a way of introducing the concept of mixed-initiative PCGML and discussing some ways these systems can be set up. We will discuss the different ways to more deeply include a user into such systems, the different interaction modalities, and what those interactions can enable within the system.

In previous chapters, we started by introducing the basics and the theory behind an approach, and then proceeded to walking through examples and exploring the approach in more depth. But as we've mentioned there is less well-established theory for this topic. As such, we will start with some discussion of existing mixed-initiative PCG tools that are used in game production, as a way of introducing the topic and providing some context for the remainder of the chapter. Then we move onto introducing the different ways to consider and enable user interactions. Next we introduce some hypothetical and prototype (i.e., not currently used in game production) mixed-initiative PCGML systems to highlight different interaction paradigms, and close with some discussion of where to go from here.

11.1 Existing PCG Tools in the Wild

There are a number of existing PCG tools that are used in game production. In this section we will first present two classical PCG tools that do not use machine learning, and then discuss two PCGML tools that have been used in actual game production. We will discuss how they create content and how users are able to interact with them.

11.1.1 Classical PCG Tools

Before we highlight existing mixed-initiative PCGML tools, it is worth taking a brief look at successful mixed-initiative PCG tools that do not rely on machine learning. One of the most well-known PCG tools is *SpeedTree* [81]. *SpeedTree* is a tool for modeling, generating, and placing trees and other foliage in games and films. While creating varied foliage may not seem especially exciting, it solves a common problem: (1) humans can often spot identical trees (or other objects) that have been placed in an environment, and (2) it is extremely time consuming for an artist to handcraft thousands and millions of similar—but not identical— trees. The tool offers the user both high-level and fine-tuned control over the content. At the high-level, the user can select templates for the trees to be used as the basis for generation, and can also indicate areas in an environment where trees should or should not be placed. For fine-tuned control, the user can directly modify the tree templates or create their own, and modify the tree placement parameters (such as frequency), or even move and modify individual trees after they've been placed. *SpeedTree* has seen wide adoption and highlights how broadly useful a mixed-initiative PCG tool can be.

Another mixed-initiative PCG tool was developed for *Horizon Zero Dawn*, an open world action RPG from 2017, where the player often explores a variety of outdoor environments. Horizon's developers created a procedural system to help the artists on the team craft expansive environments while reducing the amount of time spent hand-tuning each area [214].

This tool does not use PCGML, but a set of logic networks where each uses a combination of density maps and hand-crafted logic for determining asset placements in an environment. Overall, the system defines and leverages *ecotopes* which are tree-like structures for grouping different assets by type (e.g., the "forest" ecotope might have a "bush" child which in turn includes specific bush assets). Ecotopes are used to define (1) which objects are placed, and (2) the distribution of those objects. The user can then customize these ecotopes by defining which objects should be included, adding density maps corresponding to different objects or groups of objects, and even adding interaction rules between different density maps (e.g., subtracting a road density map from the tree density map to ensure trees are not placed on a path). From the perspective of the end users of the tool (i.e., the artist creating a particular environment), they are able to select different paint brushes corresponding to different ecotopes, objects, etc. and simply paint an area in the environment. While painting, the underlying procedural system automatically and in real time handles how different elements should be distributed within that painted region (e.g., placing different types of grass or underbrush if its painted near a rock vs. painted near the road). This system highlights the power and usability of a modular and flexible system. A more detailed description of the system can be found in the GDC talk where the system was presented [214].

It's important to also note here that many large open-world games will rely on some form of terrain or landscape generation at some point in development. For instance, *Assassins Creed Origins* [46], and *The Witcher 3* [49] both used landscape generation during development. These techniques can be used to get a starting point for the landscape which the designers can then improve and customize. With these tools in mind, we will now introduce two PCGML-powered tools.

11.1.2 Microsoft FlightSim

Microsoft FlightSim [30] is a flight simulator from Microsoft released in 2020 that includes a recreation of the world in 3D made from satellite and photogrammetry data. FlightSim uses a PCGML pipeline to create the world map and structures that the players see as they fly. This pipeline first takes in real satellite map data to initialize the terrain and scenery. Next, the satellite data is segmented and the segments classified into different groups, such as buildings, vegetation, and infrastructure. This segmentation and classification is performed by specially trained deep neural networks. A separate neural network model is then used to reconstruct the buildings, deciding the building type, height, roof, and footprint. Lastly for the buildings, facade features are added automatically based on geographical and contextual data. Photorealistic detail is achieved in some areas by incorporating data from Bing Maps, such as streetview images and other photographic data.

Unlike SpeedTree this pipeline was built with an eye towards being more self-reliant, and not requiring as much human input or feedback after the model setup. That is, this was not intended as a deeply collaborative or co-creative tool, but instead as a tool that could be setup and run somewhat autonomously. The main human inputs are in the ML model setup and in designing and creating the building blocks placed by the model (e.g., the different building pieces). However, there are still points at which a developer could intervene and fix issues in the output, either through directly fixing the output geometry (which they tend to avoid), or by updating the ML models, training behavior, and data. One common issue that requires a manual fix is the rendering of famous landmarks. Landmarks will initially be treated as any other building, and rendered with the basic design models. This can become an issue with recognizable landmarks. For example, players noticed Buckingham Palace was initially rendered as an office building; and the designers stepped in to manually fix it. To resolve this type of issue, designers create models of the landmark to be rendered and add them to the system in place of the generic models.

11.1.3 Puzzle-Maker

The last tool we will discuss is the *Puzzle Maker* [128] tool from Modl.ai. Puzzle Maker is a PCGML-based level design tool currently meant to be used by Match-3[1] level designers, and has been used by at least one Match-3 studio[2] as of writing.

The generator of the Puzzle Maker leverages a Markov random field-based level generation approaches (similar to those discussed in Chap. 6). The MRF models are trained on human-designed match-3 levels, and leverage a specialized level representation developed in collaboration with match-3 level designers. This tool follows a flow that first requests guidance from the user as to what type of levels are desired; and then uses the trained models along with some constraint enforcing/checking to generate levels adhering to the provided user inputs. Figure 11.1 shows the input interface for the Puzzle Maker, and some of the available controls (e.g., board dimension ranges, game modes, number and types of different pieces to appear, etc.). Lastly, Puzzle Maker presents the generated batches of levels along with numerous features both computed from the structure of the generated levels as well as from bot gameplay statistics. Figure 11.2 shows an example output given by the tool along with some computed features (e.g., predicted winrates, score increase per move taken, etc.). The provided evaluations are then used by the designer to select the generated levels that most closely fit their design goals; and the designer can make any final edits or modifications to the levels before deploying them. Notice that this tool allows for more user input and interaction than the FlightSim tools, but still only allows limited collaboration between the user and generator.

---

[1] For Match-3 games, think games like *Candy Crush* or *Bejeweled*.

[2] Namely, the Puzzle Maker tool was used by Goodgame Studios during the development of one of their mobile Match-3 games.

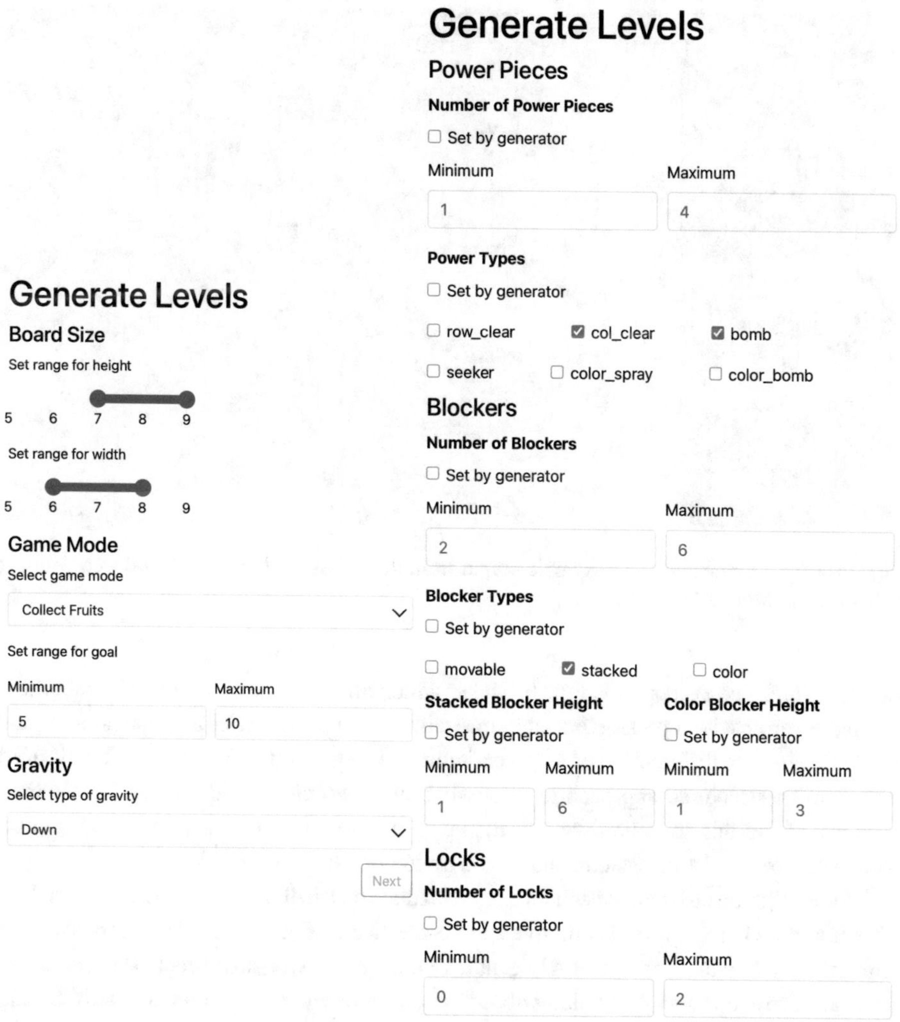

Generate Levels

Power Pieces

Number of Power Pieces

☐ Set by generator

Minimum	Maximum
1	4

Power Types

☐ Set by generator

☐ row_clear   ☑ col_clear   ☑ bomb

☐ seeker   ☐ color_spray   ☐ color_bomb

Blockers

Number of Blockers

☐ Set by generator

Minimum	Maximum
2	6

Blocker Types

☐ Set by generator

☐ movable   ☑ stacked   ☐ color

Stacked Blocker Height   **Color Blocker Height**

☐ Set by generator   ☐ Set by generator

Minimum	Maximum	Minimum	Maximum
1	6	1	3

Locks

Number of Locks

☐ Set by generator

Minimum	Maximum
0	2

Generate Levels

Board Size

Set range for height

5   6   7   8   9

Set range for width

5   6   7   8   9

Game Mode

Select game mode

Collect Fruits   ⌄

Set range for goal

Minimum	Maximum
5	10

Gravity

Select type of gravity

Down   ⌄

Next

Fig. 11.1 This figure shows available parameters for the *Puzzle Maker* tool (reproduced with permission from Modl.ai)

11.2 Structuring the Interaction

In the previous section we discussed four systems which have been useful in their respective development environments, and which offer different perspectives on user interactions with a PCG tool. The systems presented fall primarily on the two extremes of a collaboration spectrum. SpeedTree and Horizon's system enable a lot of user control and collaboration with the generative tool, while the FlightSim system only really allows for model tuning and

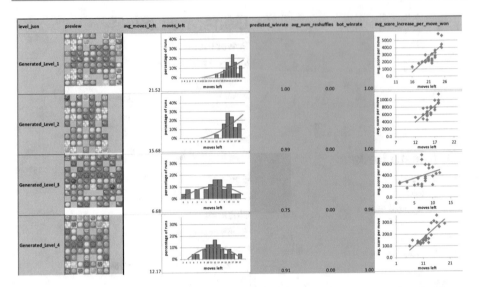

Fig. 11.2 This figure shows an example output from the *Puzzle Maker* tool (reproduced with permission from Modl.ai)

output fixing. The exception is Puzzle Maker which allows some control and collaboration through user-provided parameters and constraints. In order to better understand and more easily discuss the different types of potential mixed-initiative PCG systems, we will introduce a more structured approach to discussing and defining the user interactions in these systems. Using this structure, we will then explore more varied system setups, introduce some additional existing system, and dream up hypothetical systems.

To guide our discussion, we will leverage Guzdial and Riedl's interaction framework [62]. This framework treats mixed-initiative and co-creative PCG approaches as turn-based systems where a human user and an AI agent take turns modifying an artifact. The framework provides a structured way of talking about the different types of actions that both the user and AI can perform by categorizing them as: **artifact actions**, which directly affect the artifact being created; **other actions**, which do not directly affect the artifact; and **non-turn actions**, which the user and AI can perform external to the turn-based setup. The framework also explicitly calls out the importance of designing with a target user in mind, and with the capabilities of the AI agent in mind. Figure 11.3 shows a visualization of the framework.

11.2.1 Integrating with the PCGML Pipeline

Now that we have had a basic introduction to the interaction framework we will be referencing, we can see how that framework can be situated within a standard PCGML pipeline.

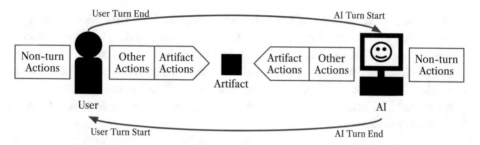

Fig. 11.3 This figure shows an illustration of the Guzdial and Riedl interaction framework [62] (recreated with permission)

Figure 11.4 (top) shows our standard PCGML pipeline. Specifically, we can see the main stages of the pipeline:

1. Data Selection
2. Model Selection and Training
3. Generation/Sampling from the Model
4. Evaluation/Characterization of the Generated Content
5. Final Content Selection.

Notice, we are using a slightly different pipeline representation than the one presented in Chap. 4. The main difference is that we split the final step in Chap. 4 ("Evaluate the Output") into two steps ("Evaluation" and "Selection"). We split this step and slightly renamed other steps in order to more easily call out where human input can be leveraged as we go through this chapter.

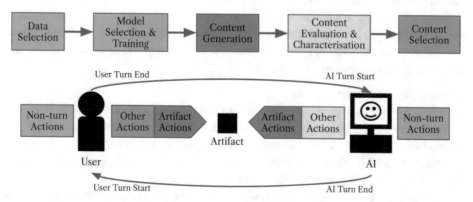

Fig. 11.4 This figure shows an illustration of a basic PCGML pipeline, flowing from training data to final content selection (top) and how categories of actions in the interaction framework could map to stages in the PCGML pipeline (bottom). Colors indicate a mapping between steps across the pipelines

With this updated pipeline, perhaps the most straightforward integration would be to simply replace the "Content Generation" node in the PCGML pipeline with the interaction framework. And in simple systems this may be enough. The data selection, model training, evaluation, and content selection can all be considered as taking place outside of the interaction framework. However, a more holistic and useful integration of the interaction framework could instead find where the different categories of user and AI actions fit within the PCGML pipeline.

Figure 11.4 shows a potential mapping between the interaction framework and a standard PCGML pipeline. To start, there may be non-turn actions by the user that relate to setting up the model and selecting training data; and similarly there may be non-turn actions by the AI which allow training or updating its models. The content generation stage of the pipeline is easily mapped to the artifact actions of both the user and the AI, since these include any actions that modify the artifact directly. The AI's other actions could include performing and presenting any kind of evaluations or characterizations of the current artifact. The user's other actions might include ending the interaction and selecting the content. Notice that the specific actions mentioned are only the ones that allow for mapping to the PCGML pipeline, and are therefore not exhaustive. Many actions of different categories could be included, such as the AI explaining its changes to the artifact as an other action or the user interacting with or playing the current artifact as a non-turn action. In order to determine which actions are possible or desirable to allow, we first need to understand the PCGML model being used and the target user of the tool.

11.2.2 Understanding the Model

When designing and building a mixed-initiative PCGML system it is important to understand the capabilities of the PCGML approach that will be used. Different models will have different affordances that enable them to perform different actions. For the remainder of the chapter we will use a hypothetical level generation tool to focus our discussions and to allow for more concrete examples. Figure 11.5 shows a mockup of an interface for this hypothetical mixed-initiative PCGML tool.

Artifact Action Considerations

Let's start with the most direct category: artifact actions. Recall that artifact actions are any actions that directly affect or modify the artifact, in this case the level being created. Artifact actions in this case could involve placing or removing elements in the level. For example, in a platformer level this might mean placing gaps or enemies, and in a dungeon crawling game this could be placing keys or treasure chests. Here we can imagine a few different setups depending on the type of model used. If we have a model that can iteratively or piece-wise generate a level (e.g., Markov chains, LSTMs, or RL models as we saw in Chaps. 6, 8, and 10), then we can set up a tool where the user and AI take turns performing artifact actions to create the final level. In our hypothetical tool in Fig. 11.5, these actions would be selecting

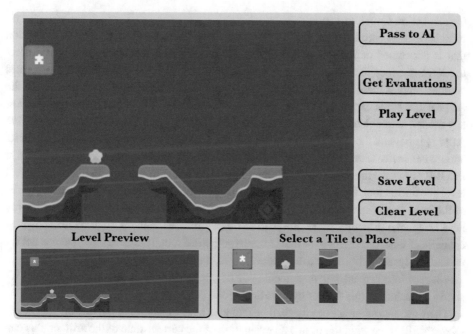

Fig. 11.5 This figure shows an example hypothetical mixed-initiative PCGML tool for creating a platformer level

a type of tile to place and then placing some of those into the level. In this scenario, the AI would take the same kind of actions once the "Pass to AI" button was pressed.

On the other hand, if the model is only set up to generate complete levels (e.g., a basic GAN approach), then the AI would fully generate a level (completing all of its artifact actions), and then the user would take their artifact actions to clean up and finalize the generated level. Again in our example system in Fig. 11.5, the level presented in the interface would have already been generated by the AI. The user could then take the actions of adding and removing specific tiles; and there would not be an option to "Pass to AI." These are somewhat specific examples, but are meant as a way of highlighting that it is important to think about how your chosen PCGML model is able to generate content, and how a user can interact with that modality.

Other Action Considerations

Next let's move onto the other actions category. Recall that other actions refer to any action that is performed during the course of the AI or user's turn that does not directly affect the artifact. Since these actions are not tied directly to how the chosen model generates content, there is more flexibility in what can be included. For example, an action where the AI provides some explanation of its updates to the artifact can potentially be included for any model; although the form of the explanation will likely need to be adapted. Explainable AI [75] is beyond the scope of this book, but the intuition is that having an AI explain its actions can increase the user's understanding of the model; and in the case of mixed-initiative PCGML, improve the user's ability to interact and design with the tool. In our hypothetical level generation tool, the AI could give a reason for why it placed an enemy in a certain area of the level (e.g., "an enemy is in this position in 95% of the training levels" or "an enemy here brings the difficulty score closer to the user-provided value"). Figure 11.6 shows an example of what this might look like in our hypothetical tool. If instead the system is not set up to directly take actions, but to suggest changes; then this is categorized as an other action. The suggestions here are still dependent upon how the model generates content. Figure 11.7 shows a mockup of this for our hypothetical tool.

There are also some actions that can be included here which might not rely on the chosen model at all! If we want to provide an evaluation of the current state of the artifact, this can

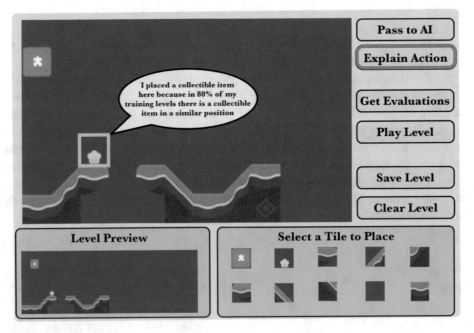

Fig. 11.6 This figure shows an example of how an explanation from the AI might look in our hypothetical tool

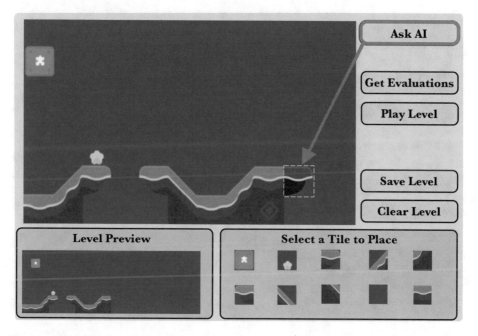

Fig. 11.7 This figure shows an example of how a suggestion from the AI might look in our hypothetical tool

be done with auxiliary functions that depend directly on the content and not the generator (e.g., structural and feature analysis of the artifact). For example, *Sentient Sketchbook* [112] is a mixed-initiative PCG tool for generating strategy game maps. The tool computes and displays evaluations of playability and resource distributions to the user as they edit the map. Figure 11.8 shows the result of a potential evaluation action in our hypothetical tool.

Non-turn Action Considerations

Lastly, let's look at non-turn actions. Recall that these are any actions that can be taken outside of the alternating turn structure. As noted in Fig. 11.4, some non-turn actions could relate to actions taken before the turn-based interactions begins, such as initial data selection and model training/tuning. However, model updates could also happen during the course of the turn-based interaction. For example, the AI agent could try to model the user behavior (e.g., from direct feedback, observed actions, etc.) and use that information to update its internal model in order to bias itself towards artifact actions that are more similar to the user's. In fact, Guzdial et al. [57] do exactly this, using active learning to update the AI's model of the user's behavior and preferences throughout the use of the tool. This user modeling and model update could be performed as a non-turn action. This type of model update and biasing will likely look different depending on the underlying PCGML approach, but should be generally achievable for most. For example, with a reinforcement learning agent we could try to transform the user feedback or action into a reward for the agent for certain

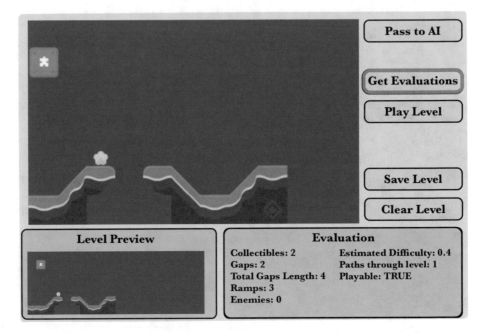

Fig. 11.8 This figure shows an example of how an evaluation from the AI might look in our hypothetical tool

actions taken; whereas for a constraint-based approach (like WaveFunctionCollapse) we could directly incorporate any new patterns observed into the allowed patterns for the model. The AI agent could also perform model updates and retraining after the interactions have finished, incorporating the newly generated artifact into its training set. Content evaluations and characterizations may be performed external to the turn-based interactions as well. After the artifact is complete, it may be useful to show how that artifact relates to the existing artifacts (e.g., in the latent space, or feature space, etc.).

11.2.3 Understanding the User

In the previous section we discussed the different action categories from the perspective of the AI agent. The target user can also have a large impact on the types of actions that should be enabled (both for the user and the AI). Who will be using the system could influence the interaction types, the granularity and format of control exposed, and the general setup of the system.

Artifact Action Considerations
Let's start again with artifact actions as they are the most direct. How the user interacts with and modifies an artifact should be tailored to their experience and expertise. Sticking with

our hypothetical level generation tool, imagine level designers of different experience levels interacting with this tool. For a more novice designer or someone new to the domain, it might make more sense for the AI agent to make larger changes, and have a tighter interaction loop, where the user makes a small update and the AI responds quickly with its own update. This setup with a tighter loop and more changes from the AI could help the user get accustomed to the domain more quickly without having the issue of "getting stuck." *Tanagra* [173] follows this flow, where the AI is continually trying to fill in the level details to satisfy the user's input. Alternatively, for a more experienced designer or someone more familiar with the domain, it may make more sense to have the AI take a more supporting role. Instead of a tight iteration loop and the AI making bigger changes, we can give control of the flow to the user. We can set up the framework such that the user indicates when their turn is over (after they've made the changes they wanted), and perhaps the AI agent only makes suggestions or small changes. The level creation tool by Guzdial et al. [57] follows this paradigm.

Other Action Considerations

In addition to the artifact actions, the user profile should also be considered when deciding which other actions to allow. As mentioned above, enabling an "ask for AI suggestion" action could be useful for users who do not necessarily want to hand over control of the design to the AI, but instead would like to see some options for moving forward (Fig. 11.6). One could also imagine a more "generate and choose" or "generate and fix" paradigm where the AI generates a complete level, and the user's only available actions are to accept or reject the artifact, or to select regions of the artifact for the AI to regenerate. The *TOAD-GAN* [155] mixed-initiative tool is set up in exactly this way, where the user is presented with a level and can request to regenerate the selected sections. The generate and choose setup could be used in scenarios where the AI is fairly robust and capable, or in scenarios where the artifact itself is very large (e.g., open world map) where a directly iterative approach does not scale well. Other sets of actions could be enabled depending on the user's expertise and knowledge of the system itself. For example, some PCGML models could be set up to work with input design parameters or constraints. *Tanagra* accepts a *beat timeline* from the user indicating the desired rhythm of the level. For a user familiar with the domain, these parameters could offer a lot of useful control over the generator (e.g., only generate match-3 levels with certain numbers of power pieces and blockers). Alternatively, directly calling out specific level examples and asking the generator for something similar is a less technical way of enabling some control as well.

Non-turn Action Considerations

Lastly, we'll discuss the non-turn actions the user can take. As we mentioned in Fig. 11.4, setting the training data and model tuning is one non-turn action that could be available to the user. However, exposing this action only really makes sense in scenarios where (1) the user has some understanding of the PCGML approach being used, (2) understands how the selected data could affect the resulting model, and (3) understands the domain enough to be able to select training data and model parameters to guide the model towards the result they want. A different non-turn action that is more broadly useful is interacting with the current version of the artifact (e.g., in Fig. 11.5, the "Play Level" button). This can allow the user to get a sense of the current state of the artifact (in terms of gameplay, and other dynamic elements) outside of just the representation being modified during the artifact actions.

11.3 Design Axes

The previous sections discussed the interaction framework, how that framework maps to a basic PCGML pipeline, and some of the considerations for setting up the interactions and actions in that framework. Now, with that context we can discuss a few potential axes of variation when setting up an interactive PCGML pipeline, as well as some examples of existing approaches that fall within these setups.

11.3.1 AI vs. User Autonomy

The first axis of variation we will discuss in these mixed-initiative systems is the axis of autonomy; that is, who has the control over the artifact during the interactions, the AI or the user?

Autonomous Generative Systems
At one end of the spectrum, we have **autonomous generative systems**. These systems most closely resemble the approaches we have discussed thus far in the book. These types of systems severely limit the actions available to the user, in favor of relying on the AI agent. In this setup, the user has very limited or no artifact actions available to them. The user may be able to set up the model (by choosing data, setting parameters, etc.) as non-turn actions, and is able to accept or reject a created artifact from artifacts created by the AI agent. However, all the artifact actions lie with the AI, which will generate entire artifacts (or batches of artifacts) to present to the user. Notice that many of the approaches discussed in previous chapters can be described with this interaction framework. One specific example of this type of tool is the GUI built for the *TOAD-GAN* approach [155]. With this tool, the user can ask the AI to generate a new level or load in a level; but the only actions the user can take to affect the level are to request the AI to modify specified level sections.

Creativity Support Systems
At the complete opposite end of the spectrum there would be no AI and it would simply be a user and an unintelligent tool. However, that is outside the purview of PCGML, so the interaction paradigm with the least amount of AI autonomy we consider are **creativity support systems**. Creativity support systems limit the AI agent's available actions in favor of relying on the user. In this paradigm the AI has very limited or no artifact actions available to it. As the user is creating the artifact, they may be able to request support from the AI in the form of suggested changes, characterizations or evaluations of the artifact, or other types

of feedback/input, but the AI cannot directly modify the artifact. *Sentient Sketchbook* [112] is an excellent example of this setup. Using *Sentient Sketchbook*, the user is in charge of modifying a map sketch directly by placing different tiles that represent game and level features for a real-time strategy map (think *StarCraft*). Meanwhile, the AI is continuously presenting design alternatives that it is generating in the background based on the user created map; as well as presenting some computed features of the current artifact (e.g., fairness of resource placements, shortest path length between bases, etc.). In this case, the AI functions as support for the user, without directly modifying the artifact itself. *Lode Encoder* [17] is another example of a support tool. With Lode Encoder, the AI creates suggestions of levels, and the user is only able to select portions of those levels to fill in the current co-created level. After changes are made, the AI will create a new set of suggested levels for the user to reference.

Co-creative Systems

Now, in between those extremes on this axis lie all other sorts of **co-creative systems**, where the user and AI share control to varying degrees. In these systems both the user and the AI will take artifact actions and contribute to the final artifact. When discussing mixed-initiative PCGML approaches, these are the types of systems that likely spring to mind. One example of such a system was already mentioned in Chap. 10, *Morai Maker* [57]. *Morai Maker* is a mixed-initiative PCGML level design tool, where the user and the AI take turns placing tiles into an empty level until the user says the level is complete. Figure 11.9 show the interface for this tool. This tool exemplifies the shared control co-creativity and turn taking for this

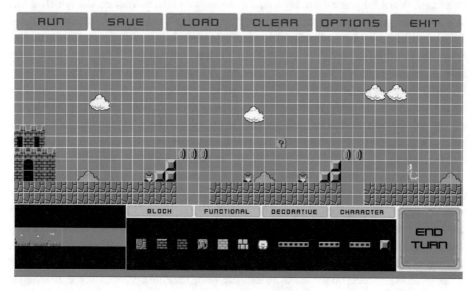

Fig. 11.9 This figure shows the editor for Guzdial et al.'s Morai Maker [57] (reproduced with permission)

region of the spectrum. However, the control, while shared, does not always have to be equal. The *Puzzle Maker* tool we discussed earlier has the AI agent generate groups of levels, and then the user can modify, select, and clean up those generated levels. In that tool the main user autonomy is in input constraints and final edits. On the flip side, there can be a system where the user creates the majority of the level, and simply uses the AI to clean up sections or fill in some details [83]. Both of these systems are considered co-creative, since they both allow the user and AI to take significant artifact actions. But notice that each of these fall further towards one direction or the other on the spectrum. Figure 11.10 shows where these hypothetical tools might fall on the spectrum relative to other tools ("User Creation with ML fixes" and "ML with user fixes"). We mention these hypothetical systems to highlight the variety and potential of how differently the interactions of mixed-initiative systems can be set up.

11.3.2 Static vs. Dynamic Model Systems

User and AI autonomy is perhaps the most apparent axis of variation for mixed-initiative PCGML systems, but another important axis is Static vs. Dynamic models. This axis is less of a spectrum and more of a binary indicating if the underlying PCGML model is updated while the tools are in use.

Static Models

On the more straightforward side of this axis, we have mixed-initiative approaches that rely on a static model for generation. By a **static** model, we mean that the mixed-initiative tool uses a specific version of a trained model over the course of a session without updating itself. These systems have the benefit of having consistent behavior across users and sessions. This setup can be useful in situations where either (1) the underlying PCGML approach is not able to be easily/quickly retrained or tuned, or (2) we are confident that the underlying PCGML model is capturing the desired design information from the training set and we do not want to move away from that. If we look at *Lode Encoder* or *TOAD-GAN* again, we can see that the models are pre-trained for the tool, and are not modified during user interactions. Instead, the variety from the tool comes from how the user interacts with the tool (i.e., asking for changes with *TOAD-GAN*, or getting several new examples with *Lode Encoder*). Similarly, if we look at a tool like *ArtBreeder* [164] an ML tool for blending different images, the underlying GAN models are not updated through interactions with the user; the variety of outputs and results come from the variety of input images provided by users. Many may assume that ML systems and tools will continue to learn and update their underlying models after they are deployed, but if we look at Fig. 11.10 we can see that the majority of mixed-initiative systems we've discussed use static models. In fact, the majority of deployed ML models do not continue learning while they are deployed unless adapting to the user is a goal of the system. Oftentimes, if additional data is collected to refine a deployed model, the model is updated or retrained offline and then redeployed. This safeguards against

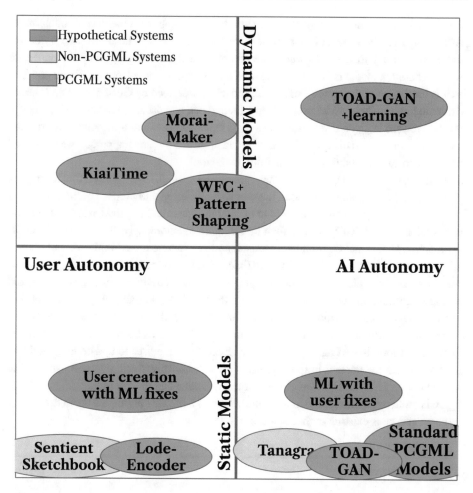

Fig. 11.10 A depiction of how the various mixed-initiative approaches we've discussed thus far compare to each other when roughly plotted on the proposed design axes of User vs. AI Autonomy and Static vs. Dynamic Models. Included are hypothetical tools we've discussed for reference. Colors are used to differentiate the hypothetical (purple), non-PCGML (yellow), and PCGML (blue) systems

model degradation either from poor or noisy data being collected or from malicious users deliberately trying to manipulate the model. In the next section we will discuss some recent work that leverages dynamic models, but it is an underexplored topic.

Dynamic Models

On the other end of the axis are mixed-initiative systems with dynamic models. A **dynamic** model in this case refers to a PCGML model that can be updated either indirectly through interactions with the user or directly by the user through parameter tuning, retraining, etc. Leveraging dynamic models opens up the ability to build tools that adapt themselves to the user's preferences, and this adaptability is their main advantage over their static counterparts. Adaptability can be a useful feature in scenarios where users will have different styles that the model can try to learn; or where the model is easily and quickly tuned/retrained; or where a baseline model is initially being used and is expected to adapt to different domains as it is used. One of the mixed-initiative approached described by Guzdial et al. [57] uses a dynamic model. Their model uses a deep RL approach (as discussed in Chap. 10), where a positive reward is given if the user keeps the agent's addition and a negative reward if the addition is removed. This type of dynamic model, mixed-initiative approach was designed explicitly to add adaptability and generality to the tool.

Karth and Smith [89] propose another dynamic mixed-initiative PCGML tool leveraging *WaveFunctionCollapse*. In their tool, the user shapes the generative space of the underlying model by providing positive examples to expand the space and negative examples to avoid undesirable content. This tool can then be run more autonomously like standard PCGML approaches. Halina and Guzdial [67] present another dynamic mixed-initiative tool for creating charts for rhythm-based games, *KiaiTime*. *KiaiTime* uses a threshold on the number of user-provided examples to determine when to retrain the underlying generative model, and during co-creation the user can specify where the agent should make changes or contributions. A final hypothetical tool might instead function more like the *TOAD-GAN* interactions, where the user can only request full levels and request sections be regenerated. But in this hypothetical tool, the AI agent could use feedback from the user to update its underlying model to make the content in those selected areas less likely. Such a system can be seen plotted in Fig. 11.10 as "*TOAD-GAN* + learning." Notice that the top (dynamic) half of the figure is quite empty compared to the bottom (static) half. This highlights again the lack of research and tools existing in that space currently.

11.4   Takeaways

In this chapter we introduced the basics behind mixed-initiative PCGML systems. We described how the basic PCGML pipeline and a mixed-initiative interaction framework can co-exist. Lastly, we discussed some of the potential design axes to consider when developing co-creative systems. As of writing, mixed-initiative and co-creative PCGML has only been explored by the research community in the last few years, and there are only a few examples of these tools being used in practice (as we saw in Sect. 11.1). Figure 11.10 shows how some existing approaches compare to each other, and more importantly highlights the large gaps in the existing research. There are large regions where no tools exist and very little research has been done (i.e., dynamic models). However, even in the limited research that has been done using mixed-initiative PCGML, there are clear advantages and clear use-cases for when co-creative systems will be preferred over fully autonomous PCGML approach.

Of course this chapter is not meant to be an exhaustive list of all the ways to build these systems. Instead, this chapter is meant to function as a starting point. Hopefully some of the tools or ideas described in this chapter sparked ideas for new tools and new paradigms. The design axes we described (AI vs. User autonomy and Static vs. Dynamic Models) are just two ways of distinguishing among existing tools. There are certainly other potential design levers to be pulled (e.g., Constructive vs. Destructive Actions, Collaborative vs. Adversarial Designers). In the following chapter we will discuss other potential avenues and open problems in PCGML; some of which relate directly to mixed-initiative PCGML and others to PCGML as a whole.

Open Problems **12**

We began this book by pointing out that the history of PCGML only began in 2013, and stating that we'd cover open problems in PCGML in this chapter. But as a matter of fact all of PCGML is an open problem. But the field is changing constantly, so it's unclear how long these problems will remain open.

In 2020, Antonios Liapis published a retrospective of the first ten years of the Workshop on Procedural Content Generation (2009–2019) [110]. In it, he plotted the kinds of approaches that were used in the PCG systems discussed in the workshop papers. While all of the others remained largely the same (search-based PCG) or decreased in popularity (grammars), ML had the most significant increase over the ten years, with the expectation that it would only continue to grow. We offer this example to indicate how difficult it is to predict future trends in the academic field of PCG, because it is itself so young. That doesn't even take into account the rapid changes in the field of ML, where cutting-edge techniques can be discarded in only a few years (see natural language processing shifting from LSTMs and GRUs to Transformers). However, we anticipate continued growth for PCGML at least in the near future. This is due in part to current fascination in the Artificial Intelligence (AI) space with ML. But ML also has the promise to lessen many of the fundamental limitations of PCG in terms of proposing general systems that can benefit the game design process.

During the past 10+ years, PCGML has already addressed a number of open problems. For example, early papers struggled to produce content that appeared similar to the original in quality. PCGML researchers have made great strides on this problem for some domains, but this introduced a swathe of new problems and reintroduced problems known to classical PCG. One of these new problems was plagiarism, how to ensure that your generator isn't just copying pieces of original content. On the other hand one of the problems shared with classical PCG was controllability, how to ensure that a generator can output content that fits some design goal (in an interactive fashion or otherwise).

© The Author(s), under exclusive license to Springer Nature Switzerland AG 2022
M. Guzdial et al., *Procedural Content Generation via Machine Learning*,
Synthesis Lectures on Games and Computational Intelligence,
https://doi.org/10.1007/978-3-031-16719-5_12

All of this being said, this book is stuck in a moment in time, largely confined to the year or so the authors spent writing it. While many of these chapters give a primer on fundamental approaches that are unlikely to change significantly in the coming years, this chapter is a bit different. Here we will attempt to cover open problems. But this almost guarantees that this chapter will be the first to become dated, as many of these open problems have seen significant progress in just the last year. Still, from our limited view here in 2022 we will try to predict: what's next for PCGML?

12.1   Identifying Open Problems

Identifying open problems is itself a difficult process. We share our approach here to help others identify other open problems or future ones in the case where all of ours are immediately solved. First, we took a more generalized look at the PCGML processes we presented in Chaps. 4 and 7. We ended up with four categories: (1) Problem Formulation, (2) Input, (3) Models/Training, and (4) Output. After identifying these categories, we considered papers published in the last few years and considered what research problem they were working on and how it, or a more abstract version, might fit into these four categories. We summarize our identified open problems in Table 12.1, and step through them in detail below.

By **Problem Formulation**, we mean the problem that we're trying to solve. These can be trying to generate levels like Super Mario Bros., music like Zelda, or creatures like Pokémon, for example. But we also use this to identify more general classes of problems, like trying to generate novel content for a game that hasn't been made yet, or trying to set up your problem so that your final generator is controllable in some way.

Input in this case means what we feed into a PCGML model, which in the majority of cases means data. In terms of data, we can think about where we're getting that data (data sources) and how we can get the most out of the data that we have (data augmentation). We also have to consider how we're choosing to represent the data. This question of representation also comes up outside of Supervised Learning, for example how we choose to represent a Markov Decision Process for Reinforcement Learning problems.

Models/Training refers to what kind of models we train, their parameters and hyperparameters, and also how we choose to train these models. PCGML has still not made use of many existing architectures. For example, no one has published work on applying a

Table 12.1 Framework for organizing open PCGML problems and twelve example problems

Problem formulation	Input	Models/Training	Output
• Underexplored content types • Novel content generation • Controllability	• Data sources • Representations • Data augmentation	• Underexplored architectures • Novel optimization approaches • Novel architectures	• Applications • Evaluation • Tools

transformer to specific game content yet, in large part due to the lack of sufficient training data to see a major difference from other sequence processing architectures. While recent work has looked into approaches similar to Diffusion-based models [132] for level design [166], there's still a lot out there to explore. We also have to consider what novel ways we might train existing architectures and what novel architectures we could develop to help improve the PCGML process. We do not cover these problems in more depth, due to lack of prior PCGML examples, and due to the fact that full coverage of this category of open problems would cover a large portion of general ML open problems.

Output includes what we do with the output from a model, but also how we can consider a trained model itself as the "output" of a PCGML process. This includes the open problem of applications, where we can actually use content output by a PCGML generator or the generator itself in a game. It also includes the open problem of how to best evaluate output content or the generator itself, and how to do so in a standardized way that allows us to better understand the performance of different models on different tasks. Finally, we can also consider tools for designers to be an open problem. But we won't discuss it further in this chapter since, due to its popularity as a research area, we spent all of Chap. 11 on it.

In the following sections we'll cover most of these identified open problems, and summarize current work (at the time of writing) that seeks to address them.

12.2 Problem Formulation

12.2.1 Underexplored Content Types

The most common type of PCGML content right now is 2D, tile-based level generation [195]. Specifically, the most, most common type of PCGML content is Mario-like levels. This has led to Mario levels being referred to as the "MNIST" of PCGML, referring to the MNIST dataset [106] which is ubiquitous in image classification and image generation research. Others have referred to Mario as the "drosophila" of PCG in reference to the famous comparison of chess to fruit flies [42]. It is helpful, for the purposes of communicating your system and for comparison's sake to make use of a standard domain like this. However, it can lead to stagnation if it's the only domain where a new system is tested.

One of the most common responses to any discussion of PCGML level generation is whether the same techniques would work for 3D. This is understandable, given that the vast majority of modern AAA games are 3D. While techniques like WaveFunctionCollapse (WFC) discussed in Chap. 5 have been applied to 3D level generation [187], the majority of PCGML techniques have not. Given that some of these techniques have been applied to 3D scene reconstruction by the graphics community [226], we can expect these techniques to apply to 3D level generation.

The primary problem for 3D PCGML level/map generation is the lack of data. Unlike 2D games where an image of a level can be translated into a tile-based representation, translating

3D game levels (even when they're available) into an appropriate representation is much more challenging. These problems, locating sufficient 3D game level data, coming up with a low-cost way to parse it into a machine learning appropriate format, and then adapting or inventing PCGML approaches to train on that data, represent significant open problems. Some recent work has looked at modifying game maps from the 3D first-person shooter Team Fortress 2 [86], but not using an ML-based approach for the modification, in part due to a lack of training data.

One other common element of games outside of levels is the rules, mechanics, or physics of the game. These are in many ways the core of the game, the code that tells the game what to do when the player presses certain inputs (e.g., Mario's jump) or what non-player entities should do when certain conditions are met. Because these things are typically represented as machine code, there's no standard, shared representation that we can use to represent the rules or mechanics of different games. This makes training an ML model to produce new mechanics difficult.

There have been attempts to develop standardized languages for games. On the academic side of things, this is typically in the service of game generation, automated game playing, or other related tasks [108, 192, 211]. These are not broad approaches meant to represent all games, and they still rely on human-authoring for defining games within them. There have been attempts to create small-scale game engines like Puzzlescript [104] and Bitsy [38], which have produced large sets of games. While it might be possible to learn from this dataset of games, no one has done so yet, and the authors again remind the reader of the importance of seeking author approval.

One approach that has seen some success has been trying to extract mechanics or mechanical information from an emulator or from the frames of a game [56, 65, 190]. However, the number of learned mechanics or mechanical affordances is typically too small to apply ML methods. Thus, while some of these learned rules have been used in a generator, this required a search-based PCG approach [61]. An alternative approach has looked at hand-annotating affordances (solid, danger, etc.) [11], but so far this has been used only to support level generation tasks and not mechanics generation directly [151, 194]. Once again there is the problem of training data for mechanics generation, but there are also related problems in terms of how to take the mechanics from many games and represent them in the same general representation. Finally, because mechanics are so varied, determining how to train an ML model on this messy data is another huge open question.

Obviously there's a lot more to games than their levels and rules, with the art, music, and stories of games all being major draws for players. These areas have all seen large ML-based generators developed outside games (for images, sounds, text, etc.). But there's been significantly less work attempting to generate these for games. For example, there has only recently been work investigating the best way to represent pixel art with DNNs [148], or combining art with game mechanics information [50]. There are also recent projects investigating how we could leverage these massive models towards the particular aesthetics of games, like finetuning GPT-2 on RPG quests [216]. We'll discuss this concept of transferring knowledge between domains (**transfer learning**) more in the next section.

The final type of content we'll discuss is the generation of whole games. This would require, at minimum, the generation of new rules and new levels, but could also require the generation of new art, animations, music, sound effects, narrative, UI, and so on. While we may be able to generate the individual parts of these things, orchestrating these parts together remains an open problem [111]. PCG researchers typically refer to generating whole games as **automated game design** [28]. This problem is relatively underexplored for PCGML. This brings us to a fundamental tension of PCGML: how do we generate content for a game that doesn't exist? Thus far, the majority of this book has focused on generating new content that is stylistically similar to existing content, but this is impossible when there is no existing content. We'll discuss some concepts for how to address this problem in the next section.

12.2.2 Novel Content Generation

The question of how to produce truly novel content with PCGML is an open one. Since supervised PCGML involves trying to learn a distribution over existing content and then sampling from that distribution we can get technically new output, but it'll still fit within the learned distribution. In other words, we can't naïvely generate Zelda dungeon rooms when we trained on Super Mario Bros. levels. This is an issue when our goal is to produce new content for a game that doesn't exist yet, such as a game in development. While we could employ PCG via Reinforcement Learning (PCGRL), writing a reward function for the new desired content would be a challenge, and it'd be impossible to base an appropriate reward function on existing data since it wouldn't yet exist.

One strategy for attempting to generate content for a non-existing game is to somehow combine or alter knowledge from existing games. A diverse array of approaches have been developed based on this strategy. They can broadly be split into two classes of approach: transfer learning or domain adaptation, and combinational creativity. In the former, a learned distribution is transformed by some mechanism, such as retraining it on out-of-distribution training data. For example, the work visualized in Fig. 12.1, in which a Markov Chain is trained on levels from Kid Icarus (top three) and Kid Kool (bottom three) and then retrained with Mario levels. This ends up producing a novel distribution that doesn't produce content like that from the original distribution, which could be used in a novel game.

The second approach commonly used to attempt to address this problem is **Combinational Creativity** (CC), also sometimes called combinatorial creativity or conceptual combination. In CC the goal is to explicitly combine two different distributions, for example by training a single model on data from different games simultaneously [152]. A secondary approach is to combine models trained on different data, recombining model parameters to produce new models without new training data. Figure 12.2 gives a visualizations of five output Bayesian networks produced by combining two other Bayesian networks according to different CC approaches, along with an example output for each.

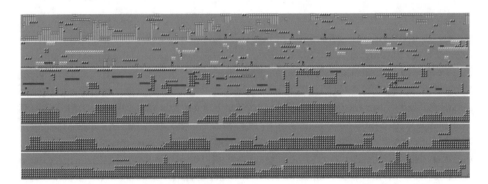

Fig. 12.1 Output levels from Domain Adaptation Markov models from Kid Icarus and Kid Kool to Super Mario Bros., reproduced with permission from [179]

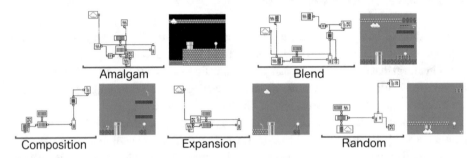

Fig. 12.2 Examples of the different final models and an output level for the different combinational creativity approaches when combining "castle" and "above ground" Super Mario Bros. levels. Reproduced with permission from [64]

Both of these approaches have thus far primarily been applied to PCGML models for level generation, however CC has been applied to classical PCG as well for all kinds of content generators, including 3D character models and narrative [137, 144]. Recently, per the writing of this book, there has been some work around extracting physics from blended, generated levels [151, 194]. While this doesn't allow for fine-grained production of game mechanics directly, there are some aspects of mechanics that can be extracted from a generated level produced via blending two different games together. For example, we can identify the jump height and distance that the player would have to traverse in the generated levels.

CC has been used to produce whole new games by recombining machine-learned models of rules and levels [63]. However, the games were of insufficient quality to be anything but curiosities. Still, if solved, this open problem could allow anyone to produce whole new games that fit their design vision. But to be able to do that we'll need to greatly improve the quality of the output content and to demonstrate the ability for these types of systems to be controllable by non-experts. Even then, this is just one potential solution for how to generate content for novel games, and this area generally could use significantly more exploration.

If we did manage to solve this problem we'd be left with many others. For example, how could we control an automated game design model to get the kind of game we wanted? This is referred to as **controllability**.

12.2.3 Controllability

Controllability refers to the problem of being able to get output content that matches some design criteria. It is related to mixed-initiative systems and creativity support tools, but refers to ways of impacting the model output outside of explicit tools. For example, recent ML image generation tools like DALL-E [143] are controllable via text prompts while tools like Artbreeder [165] are controllable via sliders. While these are tools, the backend models are separately controllable even outside of the interface.

Controlling the output of PCGML models has been a long-standing concern for the research community [40, 58, 114, 180]. However, there's not yet a sense of how best to set up models to maximize controllability. The common approach is to add extra input to the model

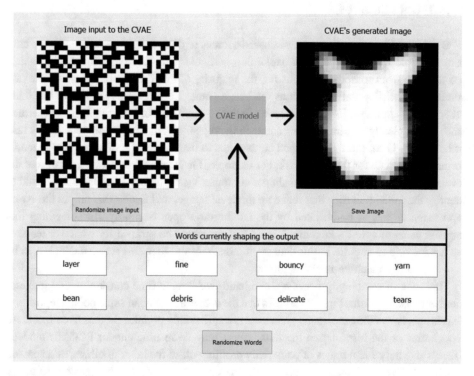

Fig. 12.3 Example of the output of a model conditioned on images and descriptions for Pokémon-like creatures. Reproduced with permission from [51]

to allow for controlling the output. For example, taking descriptions of monster silhouettes and using these to condition a model, so that changing the description changes the output [51], an interface based on this model can be seen in Fig. 12.3. Alternatively, a learned space can be searched using search-based PCG approaches, allowing users to specify preferred output with a fitness function [219]. Autoencoders and GANs are common models used in controllability research, as they allow a model to learn a single latent space that can combine multiple types of inputs, or which can be constrained or searched via other approaches.

Despite the amount of research applied to these problems, none of these systems are fully controllable, by which we indicate that it is not possible to get any of these models to output any arbitrarily specified piece of content. The open problems then remain in terms of how we improve this controllability, what kind of "controls" we give the user, and how the user can use them? One aspect of this is training data, as some of these approaches require extra training data to define the extra, "controllable" input.

12.3 Input

12.3.1 Data Sources

Outside of Chap. 3 we have not discussed sources of data, simply assuming that we have data available for the purposes of explaining particular ML methods. However, that data has to come from somewhere. Thus far, the majority of PCGML work has made use of a small set of existing datasets such as the Video Game Level Corpus [200]. We'll discuss other datasets in Chap. 13. These datasets require a good deal of human effort to construct, since they typically require scraping some online repository, most often put together by fans or hobbyists. From there the scraped content has to be translated into a machine-readable format appropriate for the target task. For example, for 2D platformer levels, images of the levels are first downloaded, then a simple computer vision library like OpenCV is used to identify the individual tiles that make up the level image, and finally this output tile-based representation has to be checked by the human developer. Similarly, for something like monster generation (e.g., Pokémon), a fan website might be scraped for whatever features we are looking to capture in our generated monsters, and then this scraped data has to be converted into a machine-readable format.

This style of scraping, processing, and hand-tweaking is time consuming and makes a number of assumptions. First, it assumes that the data is available in some pooled repository. This is not always the case, many games are not sufficiently popular for fans to collect these repositories or the information the fans collect may be insufficient for PCGML models. There's also the ethical issues of potentially profiting off of the labor of others, as discussed in Chap. 4. Even if fans have pooled existing content, it may not be possible to easily process, such as in the case of complex, game-specific files. Finally, while hand-checking the processed content is always technically possible, the scope of the task might make it infeasible. For example, if we somehow processed maps for giant open world games it may not be possible to verify every section completely.

There have been a few alternatives proposed to this process. One is to not make use of pooled repositories of game content, but to instead attempt to extract whatever content is of interest through gameplay video, video of someone playing the game in question [63]. This is a complicated process, and it leads to learning much noisier models in comparison to discrete, tile-based representations. However, this approach could potentially extend to 3D games given scene reconstruction work in the computer vision field [204]. As an alternative approach, some recent work has gone directly to the source, extracting content from emulators and ROMs of games and using PCGML to generate new content that can then be "reinjected" into the original games [122].

These alternate approaches have the potential to make significantly more training data available for PCGML. But they also have weaknesses and add extra complexity, requiring new methods to process their input into something usable by a PCGML method, in addition to new PCGML methods to use said processed input. Even the question of what representation, what form, to process this data into is an open one.

12.3.2 Representations

We have used a tile-based representation for level design across many of the chapters of this book. The strength of the tile-based representation is reflected in its popularity, but it also has clear limits. For example, tile-based representations assume the ability to represent game levels as a set of discrete, known game objects on a grid, which is not always possible. There are a number of alternatives to the standard, human-authored tile-based representation that have been explored. For example, Snodgrass attempted to automatically learn what tiles existed based on a clustering approach in tandem with a PCGML model [175]. Alternatively, Guzdial and Riedl learned a sprite-based representation of levels that is similar to the tile-based representation but did not rely on a grid [63].

One recent approach has been to learn an embedding. The concept of an embedding is that it re-represents complex, high-dimensional data in a lower dimensional space, using an autoencoder framework as discussed in Chap. 9. The work from Mawhorter et al. mentioned earlier that used a ROM of a game as input employed an embedding to translate from the raw bits and bytes of the level as it was represented in the software (high-dimensional and complex) to a discrete, small representation much closer to tiles [122]. Similar approaches have been used to learn a representation of individual entities in gameplay video [95] or to convert directly from level images to a tile-like representation [82]. Embeddings have a number of benefits, for example, they can often do a reasonable job of filling in missing information. For the Jadhav and Guzdial tile embedding work, this means that the embedding can reasonably guess the affordances of a tile from just raw pixel context—guessing which pixels represent enemies or solid game entities. However, embeddings are still relatively unexplored and so how they'll stack up in comparison to tile-based representations in terms of representational quality and fidelity is unclear.

An alternative approach to an embedding is to use a Deep Neural Network model as a representation of a game. For an embedding, an autoencoder is used to translate high dimensional content to a lower dimensional representation, but the autoencoder itself does not represent the game, just a function to transform the representation. However, in the case of approaches like World Models [65], a deep neural network (DNN) is used to represent a model of a game. This model can take as input the current frame and the player action and then return the next frame, allowing someone to essentially play a copy of the game with the DNN alone. The initial work in this area led to noisy "dream-like" recreations of games. More recent work like Nvidia's GameGAN [96], has been able to learn a higher quality recreation of the original game.

GameGAN separately models the player and the background elements, allowing the creators to swap out the player avatar and the background image with images of their choice. While this leads to different visuals, the underlying model of the game remains unchanged. There has been no attempt thus far to use this kind of World Model/GameGAN approach to produce new games or to generate content. In part this is because of the size of the learned representations. DNNs are massive, and so trying to learn a distribution over DNNs, each modeling a distinct game, would be very challenging. Still, this is a promising all-in-one way to represent existing games, with the potential to use these models to generate new games.

The lack of training data has been a consideration throughout this book. While we can't always access more training data, we can at times artificially produce more training data, a process called data augmentation.

12.3.3 Data Augmentation

Data Augmentation is typically discussed in terms of image generation or processing tasks. This is because it's fairly easy to alter things like the color of an image without it having a major impact on a human's perception of the image. This is the idea of data augmentation, we can produce new data by making alterations to our original data that does not degrade its quality. This allows us to artificially increase the size of our dataset, generally leading to higher quality final models.

If our data is image-like, for example when we use small patches of levels or where we're training on a dataset of game images, we can use some of the same types of augmentations as with computer vision approaches [163]. For example, we can horizontally flip this kind of data without changing the quality of the data in most cases. However, if it was the case that the level movement wasn't the same going left to right as right to left, or that the game images had some mechanical meaning to them impacted by the flip, this would no longer be appropriate. As a more specific example, consider Pokémon, where the elemental "type" of a monster is generally reflected in the colors of that monster's sprite. Therefore, we wouldn't want to use a data augmentation approach that changed the colors of the monster sprites, since this might impact what elemental type a player would perceive [50].

Other types of content may benefit from specific data augmentations. For example, Summerville and Mateas introduced adding player paths to levels [198]. This allows for many possible variations of a level to be produced since there are generally many possible player paths through each level. Other approaches use a secondary approach to artificially produce extra content, like using an earlier version of a PCGML model and a fitness function to find additional examples of acceptable content [210]. But there are still only a limited number of data augmentation approaches that have been explored for PCGML, and some types of content like game rules have not had any types of data augmentation approaches developed for them.

If we could find appropriate data augmentation approaches, we could increase the quality of models even for tasks with limited data available. This might then impact what models we could use, and how we could train those models to produce desired output. But even if we can improve the output quality, questions remain. What do we do with the output of a PCGML process? How do we make PCGML useful to game designers/developers and game players?

12.4 Models and Training

As a reminder, we won't cover this area of open problems in detail. If you find yourself drawn to underexplored architectures, novel optimization approaches, or novel architectures, we highly recommend exploring more machine learning-focused resources [125] and venues.

12.5 Output

12.5.1 Applications

While we imagine many potential applications of PCGML in the future, we can largely break them down into offline and online PCGML. In offline PCGML, developers and designers can employ PCGML models to help them produce the final content for a game, but the PCGML model would not be in the final game itself. In comparison, online PCGML would be where the PCGML model is in the final game, and where the player then gets to interact with it directly.

Despite the popularity of offline PCG, offline PCGML is still largely unheard of at this point. This is in part due to the lack of training data problem. If a game is currently under development then there may be very limited data to train a PCGML model, and once launched there may not be a need for new content. If we imagine that we could solve this problem, we could employ offline PCGML in all the cases where we currently employ offline PCG, such as an initial starting point for later refinement by human developers (as with *The Witcher 3*'s landscape) or to help populate an open space with unique content that is not overly critical to a game's design (as with the 3D generated trees of *Speedtree*). To fully realize this offline PCGML potential, we will require the ability to generate content that matches a desired style based on few or no examples, along with making this process accessible to those without ML expertise. Chapter 11 presented mixed-initiative tools, which may be one way to make PCGML models accessible and controllable for non-experts in ML.

A notable exception to the limited training data issue is live service games, or games that are based on having a continual stream of new content for the players to consume instead of having a contained experience. We can think of games like *Destiny*, *Halo: Infinite*, *League of Legends*, and many mobile games as fitting this category. Live service games are in an ideal position to leverage offline PCGML, because once launched there will be a starting set of training data, and there will be a constant need for new content. In fact, there have been some such collaborations with industry, for example research collaborations between PCGML researchers at Modl.ai and King (the creators of *Candy Crush*) [218]. Recall also, that the PCGML puzzle level generation tool presented in Chap. 11 was deployed in a live match-3 game from Goodgame Studios. Level designers at Goodgame Studios used the tool for creating new levels for each content update, and found that "It allows us to add a much higher amount of puzzles to the game compared to crafting them by hand. Overall, we are increasing the amount and the frequency of the content that we can release for our game."[1]

Online PCGML has slightly more examples, in large part due to WaveFunctionCollapse (WFC). WFC has been used in indie games including *Caves of Qud* [22], *Bad North* [27], and *Townscaper* [187]. Many of these games are experimental. For example, in the case of *Townscaper*, the whole purpose is to just design interesting island towns, blurring the line of game and design tool. In large part so far indie game developers have been the ones willing to put PCGML into their games, as they can afford to be experimental and take risks. Ensuring that PCGML methods are stable enough to put into larger games with more developers and more dependencies is an open research problem.

There's also the other direction, imagining what potential new types of game design might become possible with PCGML. Could we have a generator that learns to adapt to the player? Or where training the generator is in some way mechanized as part of the game design? This broaches the concept of AI-based game design, in which games are designed around interacting with a particular type of AI [212]. This is more of a design problem, but still has clear technical requirements that will likely require a PCGML expert to bring about successfully.

There's a through-line in terms of the barriers to these future applications, in that we'd first need to ensure the PCGML output was of sufficient quality. But this gets us to other open problems. What does sufficient quality mean? How do we know when we achieve it? How do we measure how far off we are?

---

[1] Quotation taken from the Modl.ai website.

12.5.2 Evaluation

In his Ph.D. Thesis, Adeel Zafar surveyed PCG evaluation approaches [223]. He found that 59% relied on secondary metrics like playability and 9% relied on expressivity analysis, an approach that we discussed in Chap. 4. The remaining 32% made use of user studies. The former, metric-based approaches suffer from the fact that they rely on different metrics or similar metrics but ones defined in different ways. This means we can't necessarily compare two sets of results directly, even if they purport to use the same metrics. Similarly, when it comes to user studies, most PCG evaluations do not directly compare different PCG generator outputs. Instead they rely on rankings or ratings (e.g., "How much do you like this piece of content?"). This leaves us with no clear way to directly compare different generators or to determine when one approach or another is better suited to a problem.

The same situation is largely true for PCGML. While there has been work that directly compares the outputs of different generators [56, 196], these are fairly unusual and generally limited to a single domain. For example, Zakaria et al. conducted an experiment comparing a large number of PCGML generation approaches for Sokoban puzzle generation [225]. While they found that LSTMs performed the best generally, this doesn't mean we should just always use LSTMs. All we know is that for Sokoban level generation LSTMs seem to outperform other approaches for certain metrics.

To grow as a field, PCGML desperately needs more cohesive, general benchmarks, allowing developers to work on common problems and to make progress towards a particular goal. Datasets like the VGLC help contribute to this in terms of giving a shared domain, but there's still disagreement in terms of how to measure success when generating levels for these tasks. Most PCGML academics treat user studies as the closest thing we have to "ground truth" or objective fact, but they are difficult to design and slow to run. The development of novel, automated processes for evaluating PCGML levels that correlate well with human measures of quality would help address many of these open problems.

12.6 Discussion

In this chapter we have overviewed a set of open problems currently facing the academic field of PCGML. As much as possible, we have tried to demonstrate how all of these problems are interconnected. By undertaking a specific research problem that addresses one of these problems, one will by necessity also address some of these other problems. This is an exciting time in PCGML research for this reason, with a great potential for impact. We'll give some pointers on where readers could get started with PCGML in the next and final chapter.

Resources and Conclusions 13

In this final chapter we cover a variety of resources in an effort to help readers get started with their own PCGML projects, and then end with some final conclusions and takeaways.

Notably, this book itself can be thought of as a resource for a reader's own PCGML project. Chapter 1 is meant as a general overview and pitch for PCGML, in case you need to convince collaborators of the utility of the approach. Chapter 2 then gives an overview of classical PCG, which may be helpful as a comparison point (why not this) or as inspiration to consider a hybrid system (classical PCG + PCGML). Chapter 3 covers the very basics of ML and some helpful keywords and concepts. Chapter 4 is perhaps the most helpful if you're about to begin a PCGML project, walking readers through the process from beginning to deployment, and discussing the ethical issues along the way. Chapters 5–10 then cover a variety of different approaches to PCGML, each with their own strengths and weaknesses. For your own PCGML project, we'd recommend picking a single approach and using the appropriate chapter as a reference point during the implementation process. Finally, Chaps. 11 and 12 discuss open problems and possible applications of PCGML, with Chap. 11 focused on one of the most popular open problems in mixed-initiative PCGML. These open problems may be helpful as inspiration or as a direct goal to shoot for.

We hope the chapters of this book will prove helpful, but by necessity of being instructional they are not as practical as they potentially could be. Therefore, we next cover specific, practical resources for learning about PCGML or undertaking a PCGML project.

13.1 PCGML Resources

The most common way to learn about PCGML would be to study at an academic institute currently pursuing PCGML research or work at a company employing and/or researching PCGML.

In terms of the academic side of things, Mark J. Nelson puts together an annual list and ranking of institutions active in technical games research https://www.kmjn.org/game-rankings/. This can be a helpful starting point in terms of finding names to reach out to in your area or specific experts. Now, "technical games research" encompasses the automated playing of games and automated generation of content (along with other topics), and only some of these individuals make use of machine learning. To differentiate, we recommend finding the academic's website (likely out of date), or Google Scholar page (which will at least have their most recent publications).

Working with an academic doesn't have to mean pursuing a degree, it can also look like a more short-term collaboration. Game AI academics are typically interested in working with industry, but will also sometimes work with hobbyists or other students if they have a relevant expertise and interest. This expertise can look like subject matter expertise, but many academics won't collaborate with you if they feel they need to teach you technical skills (unless you're one of their students). As such, it can be helpful to have example projects to point to, see **Libraries**, **Code**, and **Datasets** below.

If you feel you currently do not have relevant expertise then your options become independent study, or a more short-term education opportunity like a workshop or tutorial (sometimes hosted at an academic conference, see **Venues** below). Another short-term educational opportunity is the Game AI Summer School, run by Georgios N. Yannakakis and Julian Togelius (names you will have seen frequently in this book) https://school.gameaibook.org. Notably Sam Snodgrass, one of authors of this book, has taken part in the school as a guest lecturer.

In terms of the industry side of things, there are a number of companies actively using PCGML. Most of the major video game development/publishing houses have an in-house research team, for example, Ubisoft La Forge or EA SEED. While these groups have historically focused on advancements in graphics, more recently they have begun to focus on applications of machine learning to games [3]. There are also mobile games companies with an interest in PCGML, like Zynga and King. Modl.ai is a start-up where one of the co-authors of this book works, which has a major interest in PCGML and supporting game developers in benefiting from it. Many of these companies regularly have openings or internships. However, as above, it can be important to show you already have some experience with PCGML in order to be selected for such an opportunity.

13.1.1 Other Textbooks

We identify a few specific textbooks covering related areas that may be helpful to overview. First, Tom M. Mitchell's Machine Learning textbook [125], which does not include a thorough overview of modern Deep Neural Network (DNN) approaches, but is otherwise an excellent overview of machine learning concepts and approaches. Sutton and Barto's textbook on Reinforcement Learning is a more specific overview of this class of machine learning methods [201]. More closely related, there is Shaker et al.'s 2016 overview of classical Procedural Content Generation [158]. While this doesn't cover Machine Learning methods it does a deep dive of all the topics covered in Chap. 2.

On the industry side of things, Tanya X. Short and Tarn Adams have edited two volumes on procedural generation, "Procedural Generation in Game Design" and "Procedural Storytelling in Game Design", which are focused on these problems in the games industry [161, 162]. While not a textbook, Emily Short's blog is a well-regarded resource from an expert on narrative topics, including PCG stories, quests, and characters [1]. Similarly to Shaker et al.'s 2016 overview, these resources focus on classical Procedural Content Generation, but include more specific approaches for particular use cases.

13.1.2 Code Repositories

There are a number of PCGML repositories of code designed for reuse. One clear example of this is the WaveFunctionCollapse (WFC) (discussed in Chap. 5) repository.[1] ProcJam, a procedural content generation jam we'll discuss further below, also has a tutorial for WFC.[2] The Video Game Level Corpus [200] (discussed below) has a set of examples covering a number of different PCGML approaches.[3] Finally, the authors of this book put together a GitHub Organization[4] with example projects specifically for readers!

The Artificial Intelligence and Digital Entertainment (AIIDE) conference, also discussed further below, has an artifact evaluation process by which papers can have the systems or tools associated with them peer-reviewed. This peer-review partially evaluates the reusability and usefulness of the artifact in question. While not all of these artifacts have to do with PCGML, there are a few good examples that do. Several prior PCGML papers have used gameplay videos as a source of data, and this artifact by Luo et al. specifically covers ways to extract player data from video [118].[5] Similarly, there's "Squeezer," a tool for juicing (adding effects to) generated games, which might be helpful for making PCGML content look better after generation, but isn't directly a PCGML project (it uses classical PCG) [84].[6]

More directly PCGML, Lin et al. have their artifact "Generationmania" available on github, which is a PCGML system for producing rhythm game levels (charts) for Beatmania and similar games [115].[7] Generationmania translates input sequences of music to rhythm game charts, and so therefore it might be of interest for anyone focused on sequence-based deep neural networks as discussed in Chap. 8.

---

[1] https://github.com/mxgmn/WaveFunctionCollapse.

[2] https://www.procjam.com/tutorials/wfc/.

[3] https://github.com/TheVGLC/VGLC-Examples.

[4] https://github.com/PCGML-Book.

[5] https://github.com/icpm/experience-extraction-from-gameplay-video.

[6] https://github.com/pyjamads/Squeezer.

[7] https://github.com/xxbidiao/GenerationMania.

TOAD-GAN is a one-shot Generative Adversarial Neural Network (GAN), that has been applied to Super Mario Bros. and Mario Kart, which was developed by Schubert et al. [155]. The code for it is available, which could serve as a helpful reference point for anyone interested in applying a GAN to their PCGML problem, as discussed in Chap. 9.

For those more interested in PCGML for narrative Lin and Riedl introduced "Plug and Blend", a system that used a classical PCG approach (a simple planner) to guide the generation of a large language model towards coherent narrative [114].[8] This is an interesting example of how hybrid systems can allow for greater designer control, and another example of sequence-based deep neural networks.

Finally, Jadhav and Guzdial introduced "Tile Embeddings" as a new representation for PCGML level generation (more on the traditional representations below) [82].[9] These were mentioned in Chap. 12 as an open problem in PCGML. If solved, this could allow for general PCGML level generators. If you're more interested on the research side of things rather than an applied problem, this might be a good project to check out.

13.1.3 Libraries

The language of choice for machine learning research at the moment is Python, as can be attested from the majority of the code repositories discussed above being implemented in Python. Outside of the code repositories there are a number of libraries that can be helpful for PCGML projects.

There are a large number of libraries to help with the implementation of Deep Neural Networks (DNNs). These libraries include Tensorflow,[10] PyTorch,[11] and Keras.[12] All three are well-maintained with documentation, examples, and tutorials. If you're just getting started with DNNs, you might instead make use of TFLearn,[13] a wrapper for TensorFlow that makes working with DNNs even simpler.

Outside of DNNs, there are libraries for other traditional machine learning models such as scikit-learn,[14] a library which is itself built on other libraries typically used in ML applications including NumPy.[15]

None of these libraries are necessary for PCGML projects, and it's possible to implement ML models "from scratch," but it's not recommended. By making use of an existing library, practitioners can focus on higher level questions like the architecture and hyperparameters

---

[8] https://github.com/xxbidiao/plug-and-blend.

[9] https://github.com/js-mrunal/tile_embeddings.

[10] https://www.tensorflow.org.

[11] https://pytorch.org.

[12] https://keras.io.

[13] http://tflearn.org.

[14] https://scikit-learn.org/stable/.

[15] https://numpy.org.

of their model, rather than having to define basic functions like backpropagation. On the other hand, none of these libraries will do the work for you. Even basing your model on an existing model for a similar problem isn't always sufficient to get good results. In those cases, it's recommend to fall back on the practical and ethical considerations discussed in Chap. 4, and the model-specific concerns present in the appropriate chapter.

13.1.4 Datasets

For most PCGML projects you'll need data. While we discussed alternatives in Chaps. 4 and 12, the most common source of that data will be pre-existing datasets of video game content. For level generation projects there's the Video Game Level Corpus (VGLC) [200], which includes data for a dozen classic games in a few different formats.[16] It is by far the most popular dataset for level generation tasks. Outside of the VGLC there are a couple of datasets for specific games including Angry Birds with the Corpus for Angry Birds Level Generation[17] [224], and the "Boxoban" dataset, which is inspired by Sokoban.[18]

Outside of level generation there are datasets for other types of game content. For example, there's a dataset (and RNN) for Magic: the Gathering card generation[19] and a similar one for Yu-Gi-Oh![20] If you're interested in mechanics or affordances there's also the Videogame Affordances Corpus[21] [11]. Finally, there's the General Video Game AI Competition (GVGAI), which contains mechanics, levels, and art for very simple single-screen games [135], more on this in the next section.[22]

You'll note that this represents only a small set of content (largely levels and some mechanical information for a small number of games). Clearly missing from this list are examples of content like game music, game art like sprites or 3D models, and content like levels and mechanics for other games. Some readers will already be aware of places online, largely fan-based and unofficial, to find this kind of content. We remind those readers of the ethics of using content like this discussed in Chap. 4, but note that many of the datasets above were put together by scraping these same types of unofficial, fan-created resources. Alternatively, if you find a PCGML paper that uses a dataset that doesn't appear to be public, it's worth reaching out to the authors. In most cases, they'll be happy to share the dataset with you.

---

[16] https://github.com/TheVGLC/TheVGLC.

[17] https://github.com/AdeelZafar123/AngryBirdsDataSet.

[18] https://github.com/deepmind/boxoban-levels.

[19] https://github.com/billzorn/mtg-rnn.

[20] https://db.ygoprodeck.com/api-guide/.

[21] https://github.com/gerardrbentley/Videogame-Affordances-Corpus.

[22] http://www.gvgai.net.

13.1.5 Competitions and Jams

For some, getting involved in the PCGML community or broader PCG community might be easiest with some structure or around a particular event with a deadline. In this section we'll overview a few competitions and jams that are low-time cost and low barrier to entry, but could be a good introduction to the area and some of the people in it. As mentioned above, the GVGAI competition has a number of tracks relevant to PCGML, specifically the Level Generation Track [94] and the Rule Generation Track [93]. These tracks are not always run, but when they are they'll appear as part of the Conference on Games (COG), more on that below. Similarly, there's an Angry Birds Level Design Competition[23] and a Minecraft Settlement competition called the Generative Design in Minecraft Competition (GDMC).[24]

There are also community jams, which encourage participants to take part in an unranked PCG project over a certain time frame. Chief among these is ProcJam, a PCG jam with the tagline "make something that makes something" and which encourages experimentation and creativity.[25] ProcJam is, at this point, a significant organization with talks, tutorials, art packs, and a community zine along with its flagship jam event. If you're interested in a low stakes and accepting way into this community, our recommendation would be ProcJam. More recently the AI and Games YouTube channel (more on that in Social Media section below) hosted a Game Jam broadly about implementing games with AI in them.[26] Not all of the 100+ entries have PCG in them, but many do.

13.1.6 Venues

If you're more interested in experiencing the work of the PCGML or PCG community without necessarily having to produce work yourself, then there are particular venues where you can see PCGML in practice or research advancements.

The two places where you'll see the most PCGML (and broadly PCG) practitioner work are Roguelike Celebration[27] and the Game Developer's Conference (GDC) Game AI Summit.[28] Videos of presentations from prior years' can be found on both organization's websites. The Roguelike Celebration is specifically tailored to game developers (largely independent) working on roguelike games. Because of the nature of roguelike games, almost all of these developers make use of PCG, and because of the experimental culture of the venue, some also make use of PCGML. The GDC Game AI summit, on the other hand, is not solely focused on PCG but does include an "Experimental AI Workshop" which has highlighted

---

[23] https://aibirds.org/level-generation-competition.html.

[24] https://gendesignmc.engineering.nyu.edu.

[25] https://www.procjam.com.

[26] https://itch.io/jam/aiandgames-2021.

[27] https://roguelike.club.

[28] https://gdconf.com/summit/game-ai.

several PCGML projects including AI Dungeon 2.[29] Because GDC is a more industry-focused event, it is typically more expensive to attend, but they also eventually make most talks available through their GDC Vault.

The remaining venues we'll recommend are all academic, meaning that they focus on highlighting results from research projects, with the majority conducted in a university setting. There are exceptions to this, as there are industry research labs and some independent start-ups who also publish in these venues, but the presentations are generally less focused on public-facing applications or playable games.

Before we get into the list, a brief overview of academic venues. There are three common levels of academic venues: workshops, conferences, and journals. In computer science broadly conferences are the most common place to publish papers on the results of research. In general, the amount of research work required for a paper would be the most for a journal, the second most for a conference, and the least for a workshop. Workshops are generally places for initial feedback on new research projects, with little to no evidence of the research project having positive results. Conferences are a place to publish a paper that represents a complete, smaller research project or a substantial part of a larger one, that has gone from hypothesis all the way to results. Finally, journals (at least for computer science) tend to be a place to present a complete research project, and can be thought of as equivalent to multiple conference papers. These are not hard and fast rules, there are workshops that take more work to get into than some conferences and the same with some conferences and some journals, but this is the general breakdown of these types of venues. Workshops and conferences are annual, generally occurring at the same time every year but in different locations (when possible) and organized by different academics, with some steering committee or other group ensuring some level of consistency. Papers are submitted for inclusion for these events, undergo peer review (evaluated by fellow researchers), with some accepted and some rejected, and then at least one author of each accepted paper gives a presentation.

All the academic venues from this list are included based on the author's own experience, with some extra help from Mark J. Nelson's technical games research venues list.[30] None of these venues are solely devoted to PCGML, but all of them regularly include PCGML work. We do not include venues without a focus on games, given this book. For example, we don't describe the Musical Metacreation (MuMe) Workshop in detail, as while it focuses on research work on generating music, it is in a more general sense and not tied to games.[31] These venues are ordered from workshop to conference to journal, and then alphabetically within the group.

---

[29] https://aidungeon.cc.

[30] https://www.kmjn.org/game-rankings/.

[31] https://musicalmetacreation.org.

EXAG

The Experimental AI in Games (EXAG) Workshop is a relatively new workshop and has a wide remit, including a variety of experimental AI projects in games. Perhaps because of this, it's also one of the most popular workshops in the area and puts in effort to involve independent and small studio game developers who create unique games. For example, EXAG has hosted talks from Tanya X Short of Kitfox Games, Tarn Adams, one of the co-creators of Dwarf Fortress, and Jason Grinblat, one of the co-creators of Caves of Qud. EXAG is hosted at AIIDE, more on that below.

PCG Workshop

The Workshop on Procedural Content Generation is the longest running academic venue devoted solely to Procedural Content Generation. We mentioned it in Chap. 12 in terms of Antonios Liapis' survey of its first ten years [110]. It's where some of the earliest PCGML work was published and it continues to regularly introduce new PCGML research projects. It's hosted at FDG, more on that below.

AIIDE

The favorite conference of at least some of the authors of this book, the conference on Artificial Intelligence and Interactive Digital Entertainment (AIIDE) is the most selective of the technical games conferences. By most selective we mean that it generally has the lowest acceptance rate (number of accepted papers/number of submitted papers). This does not necessarily mean that all the papers are the best, as peer-review is not a perfect system, but it tends to mean that the papers that do get in are worth noting. AIIDE also tends to try to reach out to industry, with talks in recent years from Bioware, Ubisoft La Forge (Ubisoft's research group), and Insomniac Games. Not all the papers here are PCGML papers, but there's generally at least one session (one set of talks in between coffee breaks) on PCG.

COG

The Conference on Games (COG) was previously the conference on Computational Intelligence in Games (CIG). As the name suggests, it expanded from focusing solely on AI in games (like AIIDE or EXAG) to a more broad focus on games (like FDG below). COG is now the largest games conference, with the largest number of paper presentations. The majority of COG participants are based in institutions in Europe or South East Asia, while AIIDE tends to be more North American, and FDG tends to be North American and European. This is due to a bit of a feedback loop, because these conferences have tended to be in these locations, they tend to attract attendees from these locations, who then go on to organize the next year's conference in these locations and so on.

FDG

The conference on the Foundations of Digital Games (FDG) is a conference that includes both technical games research and games research from the humanities. This makes it an excellent venue for work that straddles both spaces, such as work that considers how PCG generators include the biases of their authors [171]. But talks range widely between liberal arts/humanities and the most technical of technical games work.

ICCC

The International Conference on Computational Creativity (ICCC) is the only one of these venues without an explicit link to games. Instead it focuses on computational creativity, the task of replicating human creativity in machines or augmenting human creativity with machines. This does mean that of these conferences, this one might touch on the widest range of media, from operas to poetry to (yes) video games. Work here tends to be more experimental when it does include PCGML, but if you find interest in the questions of what creativity and design are or what they mean through your practice, this is the academic venue for you.

ICIDS

The International Conference on Interactive Digital Storytelling (ICIDS) is a conference focused entirely on narrative; understanding, processing and generating narrative. Notably, all of these conferences are international, we include that descriptor here as its part of the acronym. ICIDS does not focus on machine learning, but research work that includes ML has increased at the conference in recent years. If you're interested in storytelling and games, this is the academic venue for you.

TOG

Transactions on Games (TOG) is the only journal listed here, as it's the only journal that regularly publishes PCGML articles. If COG and TOG sound similar, this isn't a coincidence, both are run by the Institute of Electrical and Electronics Engineers (IEEE). TOG is a venue for longer form (literally, papers get more pages to work with) work covering games research. A TOG paper often "caps" a long term technical games research project, like a Ph.D. thesis. Therefore, one can often read a single TOG article on a project that summarizes several conference and workshop papers.

13.1.7 Social Media

We'll end our group of resources with some links to where to find various PCGML sources on social media.

As mentioned previously, Tommy Thompson is the founder and host of the YouTube channel AI and Games.[32] While the channel is largely focused on covering the AI in AAA games, Tommy does have a background as an academic and so sometimes highlights specific research projects, including PCGML projects in a way meant for general public consumption. AI and Games also has a Discord as does the previously mentioned ProcJam, which you can find on their respective websites. If you prefer Facebook there's a "Game AI" group, and if you prefer no social media at all you can join the surprisingly active PCG google group email list.[33] For whatever reason Twitter seems to have become something of a haunt for

---

[32] https://www.youtube.com/channel/UCov_51F0betb6hJ6Gumxg3Q.

[33] https://groups.google.com/g/proceduralcontent.

games researchers. If you see a name associated with some interesting research, nine times out of ten you can likely find them on Twitter, including the authors of this book. We'll end our list of resources with that not too subtle suggestion to follow us on social media.

13.2   Conclusions

We thank you, anyone who has read any part of this book. This book was a challenge to write, perhaps only made possible by the fact that we three co-authors were already the first three authors on the PCGML survey paper back in 2018. We hope that we've been able to help you understand PCGML a bit better, its strengths, weaknesses, and open areas of research and exploration. We're excited about the future of PCGML, and we hope you'll join us in making it happen.

If we've succeeded even a little bit at helping foster some interest or excitement, you might be considering a PCGML project of your own. If that's the case, then we hope this book and the list of resources in this chapter specifically can serve as something like a guide or reference as you begin the project.

We do hope you'll join us as PCGML practitioners and researchers. At present, those two roles are largely the same. As we discussed in the last chapter, PCGML is not a solved field, with almost any application bound to require research in terms of developing new knowledge. However, in the future we could imagine a situation where this is not the case. In such a future, what could PCGML allow us to do?

If we somehow manage to "solve" PCGML, what does that get us? We don't imagine that PCGML will be able to replace a designer. Rather, just as with modern PCG, we imagine that PCGML will primarily take two roles: aiding designers (offline PCGML) and inside games producing content "live" (online PCGML). However, unlike classical PCG, which relies on a human developer to do a significant amount of authoring prior to deployment, a PCGML system could be much more general. Thus, we hope this added flexibility can help developers produce new kinds of experiences for players. You could imagine a game responding to player input like AI Dungeon 2, but producing levels, characters, art, music, and sound effects all at the same time. Or, unlike the conversational nature of AI Dungeon 2, a game centered around the player as designer: asking the game for an item like "this" or a boss like "that." On the developer side, we hope that the future of PCGML can not only improve the lives of existing developers but could also help broaden who gets to be a developer. A world of brand new types of games and where anyone can make a game need not be that far off. We hope you can help make this world happen.

References

1. Emily short's interactive storytelling. https://emshort.blog. Accessed: 2022-06-15.
2. Introduction to Procedural Noise. https://gdquest.mavenseed.com/lessons/introduction-to-procedural-noise, 2022. Accessed: 2022-07-06.
3. Eloi Alonso, Maxim Peter, David Goumard, and Joshua Romoff. Deep reinforcement learning for navigation in AAA video games. In Zhi-Hua Zhou, editor, *30th International Joint Conference on Artificial Intelligence, IJCAI 2021, Virtual Event / Montreal, Canada, 19-27 August 2021*, pages 2133–2139. ijcai.org, 2021. https://doi.org/10.24963/ijcai.2021/294. URL https://doi.org/10.24963/ijcai.2021/294.
4. Christophe Andrieu, Nando De Freitas, Arnaud Doucet, and Michael I Jordan. An introduction to mcmc for machine learning. *Machine learning*, 50(1):5–43, 2003.
5. Travis Archer. Procedurally generating terrain. In *44th Annual Midwest Instruction and Computing Symposium, Duluth*, pages 378–393, 2011.
6. Arul Asirvatham and Hugues Hoppe. Implementing improved Perlin noise. 2005.
7. Maren Awiszus, Frederik Schubert, and Bodo Rosenhahn. Toad-gan: coherent style level generation from a single example. In *16th AAAI Conference on Artificial Intelligence and Interactive Digital Entertainment (AIIDE)*, number 1, pages 10–16, 2020.
8. Maren Awiszus, Frederik Schubert, and Bodo Rosenhahn. World-gan: a generative model for minecraft worlds. In *3rd IEEE Conference on Games (CoG)*, pages 1–8, 2021. https://doi.org/10.1109/CoG52621.2021.9619133.
9. Dzmitry Bahdanau, Kyunghyun Cho, and Yoshua Bengio. Neural machine translation by jointly learning to align and translate. In Yoshua Bengio and Yann LeCun, editors, *1st International Conference on Learning Representations, ICLR 2015, San Diego, CA, USA, May 7-9, 2015, Conference Track Proceedings*, volume 3, 2015. URL http://arxiv.org/abs/1409.0473.
10. Breanna Baltaxe-Admony. Wfc-piano-roll. https://github.com/bbaltaxe/wfc-piano-roll, 2019.
11. Gerard R Bentley and Joseph C Osborn. The videogame affordances corpus. *6th Experimental AI in Games (EXAG) Workshop*, 2019.
12. Stefano Beretta, Mauro Castelli, Ivo Gonçalves, Roberto Henriques, and Daniele Ramazzotti. Learning the structure of bayesian networks: A quantitative assessment of the effect of different algorithmic schemes. *Complexity*, 2018, 2018.

225

M. Guzdial et al., *Procedural Content Generation via Machine Learning*, Synthesis Lectures on Games and Computational Intelligence. https://doi.org/10.1007/978-3-031-16719-5

13. Mark Berger. HeaRNNthstone: Generating Hearthstone cards with an LSTM network. https://berger-mark.medium.com/hearnnthstone-generating-hearthstone-cards-with-an-lstm-network-part-two-d98c7835aab3, 2018. Accessed: 2022-06-10.

14. Joseph Berkson. Application of the logistic function to bio-assay. *Journal of the American statistical association*, 39(227):357–365, 1944.

15. Christopher Berner, Greg Brockman, Brooke Chan, Vicki Cheung, Przemysław Dębiak, Christy Dennison, David Farhi, Quirin Fischer, Shariq Hashme, Chris Hesse, et al. Dota 2 with large scale deep reinforcement learning. *arXiv preprint* arXiv:1912.06680, 2019.

16. Dimitri Bertsekas. *Rollout, Policy Iteration, and Distributed Reinforcement Learning*. Athena Scientific, 2021.

17. Debosmita Bhaumik, Ahmed Khalifa, and Julian Togelius. Lode encoder: Ai-constrained co-creativity. In *3rd IEEE Conference on Games (CoG)*, 2021.

18. Boris The Brave. Debroglie. https://github.com/BorisTheBrave/DeBroglie, 2022.

19. Tom B. Brown, Benjamin Mann, Nick Ryder, Melanie Subbiah, Jared Kaplan, Prafulla Dhariwal, Arvind Neelakantan, Pranav Shyam, Girish Sastry, Amanda Askell, Sandhini Agarwal, Ariel Herbert-Voss, Gretchen Krueger, Tom Henighan, Rewon Child, Aditya Ramesh, Daniel M. Ziegler, Jeffrey Wu, Clemens Winter, Christopher Hesse, Mark Chen, Eric Sigler, Mateusz Litwin, Scott Gray, Benjamin Chess, Jack Clark, Christopher Berner, Sam McCandlish, Alec Radford, Ilya Sutskever, and Dario Amodei. Language models are few-shot learners. In Hugo Larochelle, Marc'Aurelio Ranzato, Raia Hadsell, Maria-Florina Balcan, and Hsuan-Tien Lin, editors, *Advances in Neural Information Processing Systems 33: Annual Conference on Neural Information Processing Systems 2020, NeurIPS 2020, December 6-12, 2020, virtual*, 2020. URL https://proceedings.neurips.cc/paper/2020/hash/1457c0d6bfcb4967418bfb8ac142f64a-Abstract.html.

20. Cameron Bolitho Browne. *Automatic Generation and Evaluation of Recombination Games*. PhD thesis, Queensland University of Technology, 2008.

21. Nathaniel Buck. *Procedural Content Generation in Strategy/Role-playing Games*. PhD thesis, University of Minnesota-Twin Cities, 2013.

22. Brian Bucklew and Jason Grinblat. Caves of qud. Freehold Games, 2015.

23. Benjamin Capps and Jacob Schrum. Using multiple generative adversarial networks to build better-connected levels for mega man. In *23rd ACM Genetic and Evolutionary Computation Conference (GECCO)*, GECCO '21, page 66-74, New York, NY, USA, 2021. Association for Computing Machinery. ISBN 9781450383509. https://doi.org/10.1145/3449639.3459323. URL https://doi.org/10.1145/3449639.3459323.

24. Eugene Chen, Christoph Sydora, Brad Burega, Anmol Mahajan, Abdullah Abdullah, Matthew Gallivan, and Matthew Guzdial. Image-to-level: Generation and repair. *Proceedings of the AAAI Conference on Artificial Intelligence and Interactive Digital Entertainment*, 16(1):189–195, Oct. 2020. URL https://ojs.aaai.org/index.php/AIIDE/article/view/7429.

25. Kyunghyun Cho, Bart Van Merriënboer, Caglar Gulcehre, Dzmitry Bahdanau, Fethi Bougares, Holger Schwenk, and Yoshua Bengio. Learning phrase representations using rnn encoder-decoder for statistical machine translation. *arXiv preprint* arXiv:1406.1078, 2014.

26. Kate Compton and Michael Mateas. Casual creators. In *6th International Conference on Computational Creativit (ICCC)*, pages 228–235, 2015.

27. Plausible Concept and Oskar Stalberg. Bad north. Raw Fury, 2018.

28. Michael Cook, Simon Colton, and Jeremy Gow. The angelina videogame design system-part i. *IEEE Transactions on Computational Intelligence and AI in Games*, 9(2):192–203, 2016.

29. Michael Cook, Jeremy Gow, Gillian Smith, and Simon Colton. Danesh: Interactive tools for understanding procedural content generators. *IEEE Transactions on Games*, pages 1–1, 2021. https://doi.org/10.1109/TG.2021.3078323.

30. Microsoft Corporation. Microsoft flight simulator. https://www.flightsimulator.com/, 2022.
31. George R Cross and Anil K Jain. Markov random field texture models. *IEEE Transactions on pattern analysis and machine intelligence*, (1):25–39, 1983.
32. George Cybenko. Approximation by superpositions of a sigmoidal function. *Mathematics of Montrol, Signals and Systems*, 2(4):303–314, 1989.
33. Steve Dahlskog, Julian Togelius, and Mark J Nelson. Linear levels through n-grams. In *18th International Academic MindTrek Conference: Media Business, Management, Content & Services*, pages 200–206, 2014.
34. Barbara De Kegel and Mads Haahr. Procedural puzzle generation: A survey. *IEEE Transactions on Games*, 12(1):21–40, 2019.
35. Omar Delarosa, Hang Dong, Mindy Ruan, Ahmed Khalifa, and Julian Togelius. Mixed-initiative level design with rl brush. In *10th International Conference on Artificial Intelligence in Music, Sound, Art and Design (EvoMUSART)*, pages 412–426. Springer, 2021.
36. Jacob Devlin, Ming-Wei Chang, Kenton Lee, and Kristina Toutanova. BERT: pre-training of deep bidirectional transformers for language understanding. In Jill Burstein, Christy Doran, and Thamar Solorio, editors, *11th Conference of the North American Chapter of the Association for Computational Linguistics: Human Language Technologies, NAACL-HLT 2019, Minneapolis, MN, USA, June 2-7, 2019, Volume 1 (Long and Short Papers)*, pages 4171–4186. Association for Computational Linguistics, 2019. https://doi.org/10.18653/v1/n19-1423. URL https://doi.org/10.18653/v1/n19-1423.
37. Joris Dormans and Sander Bakkes. Generating missions and spaces for adaptable play experiences. *IEEE Transactions on Computational Intelligence and AI in Games*, 3(3):216–228, 2011.
38. Adam Le Doux. Bitsy. https://ledoux.itch.io/bitsy, 2021.
39. Sam Earle. Using fractal neural networks to play simcity 1 and conway's game of life at variable scales. *arXiv preprint* arXiv:2002.03896, 2020.
40. Sam Earle, Maria Edwards, Ahmed Khalifa, Philip Bontrager, and Julian Togelius. Learning controllable content generators. *arXiv :2105.02993*, 2021.
41. David S Ebert, F Kenton Musgrave, Darwyn Peachey, Ken Perlin, John C Hart, and Steven Worley. *Texturing & Modeling: a Procedural Approach*. Morgan Kaufmann, 2003.
42. Nathan Ensmenger. Is chess the drosophila of artificial intelligence? a social history of an algorithm. *Social studies of science*, 42(1):5–30, 2012.
43. Julius Flimmel, Jakub Gemrot, and Vojtěch Černý. Coevolution of ai and level generators for super mario game. In *2021 IEEE Congress on Evolutionary Computation (CEC)*, pages 2093–2100. IEEE, 2021.
44. Dario Floreano, Peter Dürr, and Claudio Mattiussi. Neuroevolution: from architectures to learning. *Evolutionary Intelligence*, 1(1):47–62, 2008.
45. Matthew C Fontaine, Scott Lee, Lisa B Soros, Fernando de Mesentier Silva, Julian Togelius, and Amy K Hoover. Mapping hearthstone deck spaces through map-elites with sliding boundaries. In *21st ACM Genetic and Evolutionary Computation Conference (GECCO)*, pages 161–169, 2019.
46. Oleg Gamov. Creating the world of assassin's creed origins. *80.lv*, 2017. URL https://80.lv/articles/creating-the-world-of-assassins-creed-origins/.
47. Pablo García-Sánchez, Alberto Tonda, Giovanni Squillero, Antonio Mora, and Juan J. Merelo. Evolutionary deckbuilding in hearthstone. In *11th IEEE Conference on Computational Intelligence and Games (CIG)*, pages 1–8, 2016. https://doi.org/10.1109/CIG.2016.7860426.
48. Edoardo Giacomello, Pier Luca Lanzi, and Daniele Loiacono. Searching the latent space of a generative adversarial network to generate doom levels. In *1st IEEE Conference on Games (CoG)*, pages 1–8, 2019. https://doi.org/10.1109/CIG.2019.8848011.

49. Marcin Gollent. Landscape creation and rendering in REDengine 3. 2014 Game Developer's Conference (GDC), 2014.

50. Adrian Gonzalez, Matthew Guzdial, and Felix Ramos. Generating gameplay-relevant art assets with transfer learning. *7th Experimental AI in Games (EXAG) Workshop*, 2020.

51. Adrian Gonzalez, Matthew Guzdial, and Felix Ramos. A tool for generating monster silhouettes with a word-conditioned variational autoencoder. In *17th AAAI Conference on Artificial Intelligence and Interactive Digital Entertainment (AIIDE)*, volume 17 of *AIIDE'21*, 2021. ISBN 978-1-57735-871-8.

52. Ian J. Goodfellow, Jean Pouget-Abadie, Mehdi Mirza, Bing Xu, David Warde-Farley, Sherjil Ozair, Aaron C. Courville, and Yoshua Bengio. Generative adversarial nets. In Zoubin Ghahramani, Max Welling, Corinna Cortes, Neil D. Lawrence, and Kilian Q. Weinberger, editors, *Advances in Neural Information Processing Systems 27: Annual Conference on Neural Information Processing Systems 2014, December 8-13 2014, Montreal, Quebec, Canada*, pages 2672–2680, 2014. URL https://proceedings.neurips.cc/paper/2014/hash/5ca3e9b122f61f8f06494c97b1afccf3-Abstract.html.

53. Daniele Gravina, Ahmed Khalifa, Antonios Liapis, Julian Togelius, and Georgios N Yannakakis. Procedural content generation through quality diversity. In *21st ACM Genetic and Evolutionary Computation Conference (GECCO)*, pages 1–8. IEEE, 2019.

54. Nathan Grayson. In AI Dungeon 2, You Can Do Anything–Even Start A Rock Band Made Of Skeletons. https://kotaku.com/in-ai-dungeon-2-you-can-do-anything-even-start-a-rock-1840276553, 2019. Accessed: 2022-06-10.

55. Matthew Guzdial, Devi Acharya, Max Kreminski, Michael Cook, Mirjam Eladhari, Antonios Liapis, and Anne Sullivan. Tabletop roleplaying games as procedural content generators. *11th Workshop on Procedural Content Generation (PCG)*, pages 1–9, 2020.

56. Matthew Guzdial, Boyang Li, and Mark O. Riedl. Game engine learning from video. In Carles Sierra, editor, *26th International Joint Conference on Artificial Intelligence, IJCAI 2017, Melbourne, Australia, August 19-25, 2017*, pages 3707–3713. ijcai.org, 2017. https://doi.org/10.24963/ijcai.2017/518. URL https://doi.org/10.24963/ijcai.2017/518.

57. Matthew Guzdial, Nicholas Liao, Jonathan Chen, Shao-Yu Chen, Shukan Shah, Vishwa Shah, Joshua Reno, Gillian Smith, and Mark O Riedl. Friend, collaborator, student, manager: How design of an ai-driven game level editor affects creators. In *37th ACM Conference on Human Factors in Computing Systems (CHI)*, pages 1–13, 2019.

58. Matthew Guzdial, Joshua Reno, Jonathan Chen, Gillian Smith, and Mark Riedl. Explainable pcgml via game design patterns. *5th Experimental AI in Games (EXAG) Workshop*, 2018.

59. Matthew Guzdial and Mark Riedl. Game level generation from gameplay videos. In *12th AAAI Conference on Artificial Intelligence and Interactive Digital Entertainment (AIIDE)*, 2016a.

60. Matthew Guzdial and Mark Riedl. Learning to blend computer game levels. In *7th International Conference on Computational Creativity (ICCC)*, 2016b.

61. Matthew Guzdial and Mark Riedl. Automated game design via conceptual expansion. In *14th AAAI Conference on Artificial Intelligence and Interactive Digital Entertainment (AIIDE)*, 2018a.

62. Matthew Guzdial and Mark Riedl. An interaction framework for studying co-creative ai. *1st CHI Human-Centered Machine Learning Perspectives (HCMLP) Workshop*, 2019.

63. Matthew Guzdial and Mark O. Riedl. Conceptual game expansion. *IEEE Transactions on Games*, 14(1):93–106, 2022. https://doi.org/10.1109/TG.2021.3060005.

64. Matthew J Guzdial and Mark O Riedl. Combinatorial creativity for procedural content generation via machine learning. In *1st Knowledge Extraction from Games (KEG) Workshop*, 2018b.

65. David Ha and Jürgen Schmidhuber. World models. *arXiv preprint* arXiv:1803.10122, 2018.

66. Emily Halina and Matthew Guzdial. Taikonation: Patterning-focused chart generation for rhythm action games. In *16th International Conference on the Foundations of Digital Games (FDG) 2021*, pages 1–10, 2021.

67. Emily Halina and Matthew Guzdial. Threshold designer adaptation: Improved adaptation for designers in co-creative systems. 2022.

68. Mance E Harmon and Stephanie S Harmon. Reinforcement learning: A tutorial. 1997.

69. Erin Jonathan Hastings, Ratan K Guha, and Kenneth O Stanley. Automatic content generation in the galactic arms race video game. *IEEE Transactions on Computational Intelligence and AI in Games*, 1(4):245–263, 2009.

70. W. K. Hastings. Monte carlo sampling methods using markov chains and their applications. *Biometrika*, 57(1):97–109, 1970. ISSN 00063444. URL http://www.jstor.org/stable/2334940.

71. Martin G Helander. *Handbook of Human-Computer Interaction*. Elsevier, 2014.

72. Mark Hendrikx, Sebastiaan Meijer, Joeri Van Der Velden, and Alexandru Iosup. Procedural content generation for games: A survey. *ACM Transactions on Multimedia Computing, Communications, and Applications (TOMM)*, 9(1):1–22, 2013.

73. Curt Henrichs, Sayem Wani, and Saheen Feroz. Generating new 2d game assets using dc-gan. https://curthenrichs.github.io/CS534-Term-Project-Website, 2018.

74. Sepp Hochreiter and Jürgen Schmidhuber. Long short-term memory. *Neural Computation*, 9(8):1735–1780, 1997. https://doi.org/10.1162/neco.1997.9.8.1735. URL https://doi.org/10.1162/neco.1997.9.8.1735.

75. Andreas Holzinger. From machine learning to explainable ai. In *2018 World Symposium on Digital Intelligence for Systems and Machines (DISA)*, pages 55–66. IEEE, 2018.

76. Vincent Hom and Joe Marks. Automatic design of balanced board games. In *3rd AAAI Conference on Artificial Intelligence and Interactive Digital Entertainment (AIIDE)*, number 1, pages 25–30, 2007.

77. Amy K Hoover, Julian Togelius, and Georgios N Yannakis. Composing video game levels with music metaphors through functional scaffolding. In *1st Computational Creativity and Games (ACC) Workshop*, 2015.

78. Kurt Hornik. Approximation capabilities of multilayer feedforward networks. *Neural networks*, 4(2):251–257, 1991.

79. Lewis Horsley and Diego Perez-Liebana. Building an automatic sprite generator with deep convolutional generative adversarial networks. In *12th IEEE Conference on Computational Intelligence and Games (CIG)*, pages 134–141, 2017. https://doi.org/10.1109/CIG.2017.8080426.

80. Homan Igehy and Lucas Pereira. Image replacement through texture synthesis. In *4th IEEE International Conference on Image Processing*, volume 3, pages 186–189. IEEE, 1997.

81. Inc. Interactive Data Visualization. Speedtree. https://store.speedtree.com/, 2022.

82. Mrunal Jadhav and Matthew Guzdial. Tile embedding: A general representation for level generation. In *17th AAAI Conference on Artificial Intelligence and Interactive Digital Entertainment (AIIDE)*, pages 34–41, 2021.

83. Rishabh Jain, Aaron Isaksen, Christoffer Holmgård, and Julian Togelius. Autoencoders for level generation, repair, and recognition. In *ICCC workshop on Computational Creativity and Games (CCG)*, page 9, 2016.

84. Mads Johansen and Michael Cook. Challenges in generating juice effects for automatically designed games. In *17th AAAI Conference on Artificial Intelligence and Interactive Digital Entertainment (AIIDE)*, pages 42–49, 2021.

85. Daniel Karavolos, Antonios Liapis, and Georgios N. Yannakakis. Evolving missions for dwarf quest dungeons. In *11th IEEE Conference on Computational Intelligence and Games (CIG)*, pages 1–2, 2016. https://doi.org/10.1109/CIG.2016.7860391.

86. Konstantinos Daniel Karavolos. Orchestrating the generation of game facets via a model of gameplay. *University of Malta*, 2020.

87. Andrej Karpathy, Justin Johnson, and Li Fei-Fei. Visualizing and understanding recurrent networks. *arXiv preprint* arXiv:1506.02078, 2015.

88. Isaac Karth and Adam M Smith. Wavefunctioncollapse is constraint solving in the wild. In *12th International Conference on the Foundations of Digital Games (FDG)*, pages 1–10, 2017.

89. Isaac Karth and Adam M Smith. Addressing the fundamental tension of pcgml with discriminative learning. In *14th International Conference on the Foundations of Digital Games (FDG)*, pages 1–9, 2019.

90. Isaac Karth and Adam Marshall Smith. Wavefunctioncollapse: Content generation via constraint solving and machine learning. *IEEE Transactions on Games*, 2021.

91. Henry J Kelley. Gradient theory of optimal flight paths. *Ars Journal*, 30(10):947–954, 1960.

92. Ahmed Khalifa, Philip Bontrager, Sam Earle, and Julian Togelius. Pcgrl: Procedural content generation via reinforcement learning. In *16th AAAI Conference on Artificial Intelligence and Interactive Digital Entertainment (AIIDE)*, pages 95–101, 2020.

93. Ahmed Khalifa, Michael Cerny Green, Diego Perez-Liebana, and Julian Togelius. General video game rule generation. In *12th IEEE Conference on Computational Intelligence and Games (CIG)*, pages 170–177. IEEE, 2017.

94. Ahmed Khalifa, Diego Perez-Liebana, Simon M Lucas, and Julian Togelius. General video game level generation. In *18th ACM Genetic and Evolutionary Computation Conference (GECCO)*, pages 253–259, 2016.

95. Nazanin Yousefzadeh Khameneh and Matthew Guzdial. Entity embedding as game representation. *7th Experimental AI in Games (EXAG) Workshop*, 2020.

96. Seung Wook Kim, Yuhao Zhou, Jonah Philion, Antonio Torralba, and Sanja Fidler. Learning to simulate dynamic environments with gamegan. In *29th IEEE/CVF Conference on Computer Vision and Pattern Recognition*, pages 1231–1240, 2020.

97. Diederik P. Kingma and Max Welling. Auto-encoding variational bayes. In Yoshua Bengio and Yann LeCun, editors, *2nd International Conference on Learning Representations, ICLR 2014, Banff, AB, Canada, April 14-16, 2014, Conference Track Proceedings*, volume 2, 2014. URL http://arxiv.org/abs/1312.6114.

98. K Yu Kristen, Matthew Guzdial, and Nathan Sturtevant. The definition-context-purpose paradigm and other insights from industry professionals about the definition of a quest. In *17th AAAI Conference on Artificial Intelligence and Interactive Digital Entertainment (AIIDE)*, volume 17, pages 107–114, 2021.

99. S. Kullback and R. A. Leibler. On Information and Sufficiency. *The Annals of Mathematical Statistics*, 22(1):79 – 86, 1951. https://doi.org/10.1214/aoms/1177729694. URL https://doi.org/10.1214/aoms/1177729694.

100. Hans Kunsch, Stuart Geman, and Athanasios Kehagias. Hidden markov random fields. *The Annals of Applied Probability*, 5(3):577–602, 1995.

101. Ares Lagae, Sylvain Lefebvre, Rob Cook, Tony DeRose, George Drettakis, David S Ebert, John P Lewis, Ken Perlin, and Matthias Zwicker. A survey of procedural noise functions. In *Computer Graphics Forum*, volume 29, pages 2579–2600. Wiley Online Library, 2010.

102. Ares Lagae, Sylvain Lefebvre, George Drettakis, and Philip Dutré. Procedural noise using sparse gabor convolution. *ACM Transactions on Graphics (TOG)*, 28(3):1–10, 2009.

103. Gorm Lai, Frederic Fol Leymarie, and William Latham. On mixed-initiative content creation for video games. *IEEE Transactions on Games*, 2022.

104. Stephen Lavelle. Puzzlescript. https://www.puzzlescript.net, 2014.

105. Conor Lazarou. I generated thousands of new pokemon using ai. https://towardsdatascience.com/i-generated-thousands-of-new-pokemon-using-ai-f8f09dc6477e, 2020.

106. Yann LeCun. The mnist database of handwritten digits. *http://yann. lecun. com/exdb/mnist/*, 1998.

107. Scott Lee, Aaron Isaksen, Christoffer Holmgård, and Julian Togelius. Predicting resource locations in game maps using deep convolutional neural networks. In *12th AAAI Conference on Artificial Intelligence and Interactive Digital Entertainment (AIIDE)*, 2016.

108. John Levine, Clare Bates Congdon, Marc Ebner, Graham Kendall, Simon M Lucas, Risto Miikkulainen, Tom Schaul, and Tommy Thompson. General video game playing. 2013.

109. Yubin Liang, Wanxiang Li, and Kokolo Ikeda. Procedural content generation of rhythm games using deep learning methods. In Erik D. Van der Spek, Stefan Göbel, Ellen Yi-Luen Do, Esteban Clua, and Jannicke Baalsrud Hauge, editors, *Entertainment Computing and Serious Games - First IFIP TC 14 Joint International Conference, ICEC-JCSG 2019, Arequipa, Peru, November 11-15, 2019, Proceedings*, volume 11863 of *Lecture Notes in Computer Science*, pages 134–145. Springer, 2019. https://doi.org/10.1007/978-3-030-34644-7_11. URL https://doi.org/10.1007/978-3-030-34644-7_11.

110. Antonios Liapis. 10 years of the pcg workshop: Past and future trends. *11th Workshop on Procedural Content Generation (PCG)*, pages 1–10, 2020.

111. Antonios Liapis, Georgios N Yannakakis, Mark J Nelson, Mike Preuss, and Rafael Bidarra. Orchestrating game generation. *IEEE Transactions on Games*, 11(1):48–68, 2018.

112. Antonios Liapis, Georgios N Yannakakis, Julian Togelius, et al. Sentient sketchbook: Computer-aided game level authoring. In *8th International Conference on the Foundations of Digital Games (FDG)*, pages 213–220, 2013.

113. George James Lidstone. Note on the general case of the bayes-laplace formula for inductive or a posteriori probabilities. *Transactions of the Faculty of Actuaries*, 8(182-192):13, 1920.

114. Zhiyu Lin and Mark O Riedl. Plug-and-blend: A framework for plug-and-play controllable story generation with sketches. In *17th AAAI Conference on Artificial Intelligence and Interactive Digital Entertainment (AIIDE)*, pages 58–65, 2021.

115. Zhiyu Lin, Kyle Xiao, and Mark Riedl. Generationmania: Learning to semantically choreograph. In *15th AAAI Conference on Artificial Intelligence and Interactive Digital Entertainment (AIIDE)*, pages 52–58, 2019.

116. Jialin Liu, Sam Snodgrass, Ahmed Khalifa, Sebastian Risi, Georgios N Yannakakis, and Julian Togelius. Deep learning for procedural content generation. *Neural Computing and Applications*, 33(1):19–37, 2021.

117. Yinhan Liu, Myle Ott, Naman Goyal, Jingfei Du, Mandar Joshi, Danqi Chen, Omer Levy, Mike Lewis, Luke Zettlemoyer, and Veselin Stoyanov. Roberta: A robustly optimized bert pretraining approach. *arXiv preprint* arXiv:1907.11692, 2019.

118. Zijin Luo, Matthew Guzdial, Nicholas Liao, and Mark Riedl. Player experience extraction from gameplay video. In *14th AAAI Conference on Artificial Intelligence and Interactive Digital Entertainment (AIIDE)*, 2018.

119. Athar Mahmoudi-Nejad, Matthew Guzdial, and Pierre Boulanger. Arachnophobia exposure therapy using experience-driven procedural content generation via reinforcement learning (edpcgrl). In *17th AAAI Conference on Artificial Intelligence and Interactive Digital Entertainment*, pages 164–171, 2021.

120. Andrey Andreyevich Markov. Extension of the limit theorems of probability theory to a sum of variables connected in a chain. *The Notes of the Imperial Academy of Sciences of St. Petersburg VIII Series*.

121. Andrey Andreyevich Markov. Extension of the law of large numbers to dependent quantities. *Izv. Fiz.-Matem. Obsch. Kazan Univ.(2nd Ser)*, 15(1):135–156, 1906.

122. Ross Mawhorter, Batu Aytemiz, Isaac Karth, and Adam Smith. Content reinjection for super metroid. In *17th AAAI Conference on Artificial Intelligence and Interactive Digital Entertainment (AIIDE)*, pages 172–178, 2021.

123. Paul C Merrell. *Model Synthesis*. PhD thesis, The University of North Carolina at Chapel Hill, 2009.

124. Reed Milewicz. The mnist database of handwritten digits. https://www.mtgsalvation.com/forums/magic-fundamentals/custom-card-creation/612057-generating-magic-cards-using-deep-recurrent-neural, 2015.

125. TM Mitchell. Machine learning, mcgraw-hill higher education. *New York*, 1997.

126. Volodymyr Mnih, Adrià Puigdomènech Badia, Mehdi Mirza, Alex Graves, Timothy P. Lillicrap, Tim Harley, David Silver, and Koray Kavukcuoglu. Asynchronous methods for deep reinforcement learning. In Maria-Florina Balcan and Kilian Q. Weinberger, editors, *33rd International Conference on Machine Learning, ICML 2016, New York City, NY, USA, June 19-24, 2016*, volume 48 of *JMLR Workshop and Conference Proceedings*, pages 1928–1937. JMLR.org, 2016. URL http://proceedings.mlr.press/v48/mniha16.html.

127. Volodymyr Mnih, Koray Kavukcuoglu, David Silver, Alex Graves, Ioannis Antonoglou, Daan Wierstra, and Martin Riedmiller. Playing atari with deep reinforcement learning. *arXiv preprint* arXiv:1312.5602, 2013.

128. Modl.ai. Puzzle maker. https://modl.ai/our_products/modl-create/, 2022.

129. Jean-Baptiste Mouret and Jeff Clune. Illuminating search spaces by mapping elites. *arXiv preprint* arXiv:1504.04909, 2015.

130. Sanggyu Nam, Chu-Hsuan Hsueh, and Kokolo Ikeda. Generation of game stages with quality and diversity by reinforcement learning in turn-based rpg. *IEEE Transactions on Games*, pages 1–1, 2021. https://doi.org/10.1109/TG.2021.3113313.

131. Hermann Ney, Ute Essen, and Reinhard Kneser. On structuring probabilistic dependences in stochastic language modelling. *Comput. Speech Lang.*, 8(1):1–38, 1994. https://doi.org/10.1006/csla.1994.1001. URL https://doi.org/10.1006/csla.1994.1001.

132. Alexander Quinn Nichol, Prafulla Dhariwal, Aditya Ramesh, Pranav Shyam, Pamela Mishkin, Bob McGrew, Ilya Sutskever, and Mark Chen. GLIDE: towards photorealistic image generation and editing with text-guided diffusion models. In Kamalika Chaudhuri, Stefanie Jegelka, Le Song, Csaba Szepesvári, Gang Niu, and Sivan Sabato, editors, *39th International Conference on Machine Learning, ICML 2022, 17-23 July 2022, Baltimore, Maryland, USA*, volume 162 of *Proceedings of Machine Learning Research*, pages 16784–16804. PMLR, 2022. URL https://proceedings.mlr.press/v162/nichol22a.html.

133. Martin O'Leary. oisin. https://github.com/mewo2/oisin, 2017.

134. Niki Parmar, Ashish Vaswani, Jakob Uszkoreit, Lukasz Kaiser, Noam Shazeer, Alexander Ku, and Dustin Tran. Image transformer. In Jennifer Dy and Andreas Krause, editors, *Proceedings of the 35th International Conference on Machine Learning*, volume 80 of *Proceedings of Machine Learning Research*, pages 4055–4064. PMLR, 10–15 Jul 2018. URL https://proceedings.mlr.press/v80/parmar18a.html.

135. Diego Perez-Liebana, Spyridon Samothrakis, Julian Togelius, Tom Schaul, and Simon M Lucas. General video game ai: Competition, challenges and opportunities. In *30th AAAI Conference on Artificial Intelligence*, 2016.

136. Ken Perlin. An image synthesizer. *ACM Siggraph Computer Graphics*, 19(3):287–296, 1985.

137. Justin Permar and Brian Magerko. A conceptual blending approach to the generation of cognitive scripts for interactive narrative. In *9th AAAI Conference on Artificial Intelligence and Interactive Digital Entertainment (AIIDE)*, 2013.

138. Przemyslaw Prusinkiewicz and Aristid Lindenmayer. *The Algorithmic Beauty of Plants*. Springer Science & Business Media, 2012.

139. Christopher Purdy, Xinyu Wang, Larry He, and Mark Riedl. Predicting generated story quality with quantitative measures. In *14th AAAI Conference on Artificial Intelligence and Interactive Digital Entertainment (AIIDE)*, 2018.

140. Lara Raad, Axel Davy, Agnès Desolneux, and Jean-Michel Morel. A survey of exemplar-based texture synthesis. *Annals of Mathematical Sciences and Applications*, 3(1):89–148, 2018.

141. Lawrence Rabiner and Biinghwang Juang. An introduction to hidden markov models. *IEEE ASSP magazine*, 3(1):4–16, 1986.

142. Alec Radford, Jeffrey Wu, Rewon Child, David Luan, Dario Amodei, Ilya Sutskever, et al. Language models are unsupervised multitask learners. *OpenAI blog*, 1(8):9, 2019.

143. Aditya Ramesh, Mikhail Pavlov, Gabriel Goh, Scott Gray, Chelsea Voss, Alec Radford, Mark Chen, and Ilya Sutskever. Zero-shot text-to-image generation. In *38th International Conference on Machine Learning (ICML)*, pages 8821–8831. PMLR, 2021.

144. Paulo Ribeiro, Francisco C Pereira, Bruno F Marques, Bruno Leitão, Amílcar Cardoso, II Polo, and Pinhal de Marrocos. A model for creativity in creature generation. In *4th European GAME-ON Conference*, page 175, 2003.

145. Sebastian Risi, Joel Lehman, David B D'Ambrosio, Ryan Hall, and Kenneth O Stanley. Petalz: Search-based procedural content generation for the casual gamer. *IEEE Transactions on Computational Intelligence and AI in Games*, 8(3):244–255, 2015.

146. David E Rumelhart, Geoffrey E Hinton, and Ronald J Williams. Learning representations by vack-propagating errors. *nature*, 323(6088):533–536, 1986.

147. Shibani Santurkar, Dimitris Tsipras, Andrew Ilyas, and Aleksander Madry. How does batch normalization help optimization? In Samy Bengio, Hanna M. Wallach, Hugo Larochelle, Kristen Grauman, Nicolò Cesa-Bianchi, and Roman Garnett, editors, *Advances in Neural Information Processing Systems 31: Annual Conference on Neural Information Processing Systems 2018, NeurIPS 2018, December 3-8, 2018, Montréal, Canada*, pages 2488–2498, 2018. URL https://proceedings.neurips.cc/paper/2018/hash/905056c1ac1dad141560467e0a99e1cf-Abstract.html.

148. Akash Saravanan and Matthew Guzdial. Pixel vq-vaes for improved pixel art representation. *arXiv preprint* arXiv:2203.12130, 2022.

149. Anurag Sarkar and Seth Cooper. Blending levels from different games using lstms. *5th Experimental AI in Games (EXAG) Workshop*, 2018.

150. Anurag Sarkar and Seth Cooper. Towards game design via creative machine learning (gdcml). In *2nd IEEE Conference on Games (CoG)*, pages 744–751. IEEE, 2020.

151. Anurag Sarkar, Adam Summerville, Sam Snodgrass, Gerard Bentley, and Joseph Osborn. Exploring level blending across platformers via paths and affordances. In *16th AAAI Conference on Artificial Intelligence and Interactive Digital Entertainment (AIIDE)*, volume 16, pages 280–286, 2020a.

152. Anurag Sarkar, Zhihan Yang, and Seth Cooper. Controllable level blending between games using variational autoencoders. *7th Experimental AI in Games (EXAG) Workshop*, 2020b.

153. Tom Schaul, John Quan, Ioannis Antonoglou, and David Silver. Prioritized experience replay. In Yoshua Bengio and Yann LeCun, editors, *4th International Conference on Learning Representations, ICLR 2016, San Juan, Puerto Rico, May 2-4, 2016, Conference Track Proceedings*, 2016. URL http://arxiv.org/abs/1511.05952.

154. Jacob Schrum, Jake Gutierrez, Vanessa Volz, Jialin Liu, Simon Lucas, and Sebastian Risi. Interactive evolution and exploration within latent level-design space of generative adversarial networks. In *22nd ACM Genetic and Evolutionary Computation Conference (GECCO)*, pages 148–156, 2020.

155. Frederik Schubert, Maren Awiszus, and Bodo Rosenhahn. Toad-gan: A flexible framework for few-shot level generation in token-based games. *IEEE Transactions on Games*, 14(2):284–293, 2022. https://doi.org/10.1109/TG.2021.3069833.

156. Ygor Rebouças Serpa and Maria Andréia Formico Rodrigues. Towards machine-learning assisted asset generation for games: a study on pixel art sprite sheets. In *18th Brazilian Symposium on Computer Games and Digital Entertainment (SBGames)*, pages 182–191. IEEE, 2019.

157. N. Shaker, J. Togelius, G. N. Yannakakis, B. Weber, T. Shimizu, T. Hashiyama, N. Sorenson, P. Pasquier, P. Mawhorter, G. Takahashi, G. Smith, and R. Baumgarten. The 2010 mario ai championship: Level generation track. *IEEE Transactions on Computational Intelligence and AI in Games*, 3(4):332–347, 2011. https://doi.org/10.1109/TCIAIG.2011.2166267.

158. Noor Shaker, Antonios Liapis, Julian Togelius, Ricardo Lopes, and Rafael Bidarra. Constructive generation methods for dungeons and levels. In Noor Shaker, Julian Togelius, and Mark J. Nelson, editors, *Procedural Content Generation in Games: A Textbook and an Overview of Current Research*, pages 31–55. Springer, 2016a.

159. Noor Shaker, Miguel Nicolau, Georgios N Yannakakis, Julian Togelius, and Michael O'neill. Evolving levels for super mario bros using grammatical evolution. In *7th IEEE Conference on Computational Intelligence and Games (CIG)*, pages 304–311. IEEE, 2012.

160. Noor Shaker, Julian Togelius, and Mark J Nelson. *Procedural Content Generation in Games*. Springer, 2016b.

161. Tanya Short and Tarn Adams. *Procedural Generation in Game Design*. CRC Press, 2017.

162. Tanya X Short and Tarn Adams. *Procedural Storytelling in Game Design*. CRC Press, 2019.

163. Connor Shorten and Taghi M Khoshgoftaar. A survey on image data augmentation for deep learning. *Journal of Big Data*, 6(1):1–48, 2019.

164. Joel Simon. ArtBreeder. https://artbreeder.com/, 2018. Accessed: 2021-11-01.

165. Joel Simon. Artbreeder. https://artbreeder.com, 2020.

166. Matthew Siper, Ahmed Khalifa, and Julian Togelius. Path of destruction: Learning an iterative level generator using a small dataset. *arXiv preprint* arXiv:2202.10184, 2022.

167. Ruben Smelik, Krzysztof Galka, Klaas Jan De Kraker, Frido Kuijper, and Rafael Bidarra. Semantic constraints for procedural generation of virtual worlds. In *2nd International Workshop on Procedural Content Generation in Games*, pages 1–4, 2011.

168. Ruben M Smelik, Tim Tutenel, Rafael Bidarra, and Bedrich Benes. A survey on procedural modelling for virtual worlds. In *Computer Graphics Forum*, volume 33, pages 31–50. Wiley Online Library, 2014.

169. Adam M Smith, Eric Butler, and Zoran Popovic. Quantifying over play: Constraining undesirable solutions in puzzle design. In *8th International Conference on the Foundations of Digital Games (FDG)*, pages 221–228, 2013.

170. Adam M Smith and Michael Mateas. Answer set programming for procedural content generation: A design space approach. *IEEE Transactions on Computational Intelligence and AI in Games*, 3(3):187–200, 2011.

171. Gillian Smith. What do we value in procedural content generation? In *12th International Conference on the Foundations of Digital Games*, pages 1–2, 2017.

172. Gillian Smith and Jim Whitehead. Analyzing the expressive range of a level generator. *1st Workshop on Procedural Content Generation (PCG)*, pages 1–7, 2010.

173. Gillian Smith, Jim Whitehead, and Michael Mateas. Tanagra: Reactive planning and constraint solving for mixed-initiative level design. *IEEE Transactions on Computational Intelligence and AI in Games*, 3(3):201–215, 2011.

174. Gillian Smith, Jim Whitehead, Michael Mateas, Mike Treanor, Jameka March, and Mee Cha. Launchpad: A rhythm-based level generator for 2-d platformers. *IEEE Transactions on Computational Intelligence and AI in Games*, 3(1):1–16, 2010.

175. Sam Snodgrass. Towards automatic extraction of tile types from level images. *5th Experimental AI in Games (EXAG) Workshop*, 2018.

176. Sam Snodgrass and Santiago Ontañón. Generating maps using markov chains. In *9th AAAI Conference on Artificial Intelligence and Interactive Digital Entertainment Conference (AIIDE)*, 2013.

177. Sam Snodgrass and Santiago Ontañón. Experiments in map generation using Markov chains. In *9th International Conference on the Foundations of Digital Games (FDG)*, 2014.

178. Sam Snodgrass and Santiago Ontañón. A hierarchical MdMC approach to 2D video game map generation. In *11th AAAI Conference on Artificial Intelligence and Interactive Digital Entertainment (AIIDE)*, volume 11, 2015.

179. Sam Snodgrass and Santiago Ontañón. An approach to domain transfer in procedural content generation of two-dimensional videogame levels. In *12th AAAI Conference on Artificial Intelligence and Interactive Digital Entertainment (AIIDE)*, 2016.

180. Sam Snodgrass and Santiago Ontañón. Controllable procedural content generation via constrained multi-dimensional markov chain sampling. In Subbarao Kambhampati, editor, *25th International Joint Conference on Artificial Intelligence, IJCAI 2016, New York, NY, USA, 9-15 July 2016*, pages 780–786. IJCAI/AAAI Press, 2016. URL http://www.ijcai.org/Abstract/16/116.

181. Sam Snodgrass and Santiago Ontañón. Learning to generate video game maps using Markov models. *IEEE Transactions on Computational Intelligence and AI in Games*, 9(4):410–422, 2016.

182. Sam Snodgrass and Santiago Ontañón. Procedural level generation using multi-layer level representations with mdmcs. In *12th IEEE Conference on Computational Intelligence and Games (CIG)*, pages 280–287. IEEE, 2017.

183. Sam Snodgrass and Anurag Sarkar. Multi-domain level generation and blending with sketches via example-driven bsp and variational autoencoders. In *15th International Conference on the Foundations of Digital Games (FDG)*, FDG '20, New York, NY, USA, 2020. Association for Computing Machinery. ISBN 9781450388078. https://doi.org/10.1145/3402942.3402948. URL https://doi.org/10.1145/3402942.3402948.

184. Irwin Sobel et al. An isotropic 3x3 image gradient operator. *Presentation at Stanford AI Project*, 2014(02), 1968.

185. Jacob Sobolev. How many video games exist? *Remarkable Coder*. URL https://remarkablecoder.com/how-many-video-games-exist/.

186. Terence Soule, Samantha Heck, Thomas E. Haynes, Nicholas Wood, and Barrie D. Robison. Darwin's demons: Does evolution improve the game? In Giovanni Squillero and Kevin Sim, editors, *Applications of Evolutionary Computation - 20th European Conference, EvoApplications 2017, Amsterdam, The Netherlands, April 19-21, 2017, Proceedings, Part I*, volume 10199 of *Lecture Notes in Computer Science*, pages 435–451, 2017. https://doi.org/10.1007/978-3-319-55849-3_29. URL https://doi.org/10.1007/978-3-319-55849-3_29.

187. Oskar Stalberg. Townscaper. Steam, 2020.

188. Nathan Sturtevant, Nicolas Decroocq, Aaron Tripodi, and Matthew Guzdial. The unexpected consequence of incremental design changes. In *16th AAAI Conference on Artificial Intelligence and Interactive Digital Entertainment (AIIDE)*, pages 130–136, 2020.

189. Adam Summerville. Expanding expressive range: Evaluation methodologies for procedural content generation. In *14th AAAI Conference on Artificial Intelligence and Interactive Digital Entertainment (AIIDE)*, 2018a.

190. Adam Summerville, Morteza Behrooz, Michael Mateas, and Arnav Jhala. What does that ?-block do? learning latent causal affordances from mario play traces. In *Workshops at the 31st AAAI Conference on Artificial Intelligence*, 2017.

191. Adam Summerville, Matthew Guzdial, Michael Mateas, and Mark O Riedl. Learning player tailored content from observation: Platformer level generation from video traces using lstms. *4th Experimental AI in Games (EXAG) Workshop*, 2016.

192. Adam Summerville, Chris Martens, Ben Samuel, Joseph Osborn, Noah Wardrip-Fruin, and Michael Mateas. Gemini: Bidirectional generation and analysis of games via asp. In *14th AAAI Conference on Artificial Intelligence and Interactive Digital Entertainment (AIIDE)*, volume 14, pages 123–129, 2018a.

193. Adam Summerville and Michael Mateas. Sampling hyrule: Sampling probabilistic machine learning for level generation. In *11th AAAI Conference on Artificial Intelligence and Interactive Digital (AIIDE)*, 2015.

194. Adam Summerville, Anurag Sarkar, Sam Snodgrass, and Joseph C Osborn. Extracting physics from blended platformer game levels. *7th Experimental AI in Games (EXAG) Workshop*, 2020.

195. Adam Summerville, Sam Snodgrass, Matthew Guzdial, Christoffer Holmgård, Amy K Hoover, Aaron Isaksen, Andy Nealen, and Julian Togelius. Procedural content generation via machine learning (pcgml). *IEEE Transactions on Games*, 10(3):257–270, 2018b.

196. Adam J Summerville. *Learning from Games for Generative Purposes*. University of California, Santa Cruz, 2018b.

197. Adam J Summerville, Morteza Behrooz, Michael Mateas, and Arnav Jhala. The learning of zelda: Data-driven learning of level topology. *6th Workshop on Procedural Content Generation (PCG)*, 2015.

198. Adam J Summerville and Michael Mateas. Super mario as a string: Platformer level generation via lstms. *1st Joint DIGRA/FDG Conference*, 2016a.

199. Adam James Summerville and Michael Mateas. Mystical tutor: A magic: The gathering design assistant via denoising sequence-to-sequence learning. In *12th AAAI Conference on Artificial Intelligence and Interactive Digital Entertainment*, 2016b.

200. Adam James Summerville, Sam Snodgrass, Michael Mateas, and Ontañón. The vglc: The video game level corpus.

201. Richard S Sutton and Andrew Barto. *Reinforcement learning: an introduction*. The MIT Press, 2 edition, 2020.

202. Keijiro Takahashi. Wfcmaze. https://github.com/keijiro/WfcMaze, 2021.

203. Takumi Tanabe, Kazuto Fukuchi, Jun Sakuma, and Youhei Akimoto. Level generation for angry birds with sequential vae and latent variable evolution. In *23rd ACM Genetic and Evolutionary Computation Conference (GECCO)*, pages 1052–1060, 2021.

204. Zachary Teed and Jia Deng. Raft-3d: Scene flow using rigid-motion embeddings. In *30th IEEE/CVF Conference on Computer Vision and Pattern Recognition*, pages 8375–8384, 2021.

205. Tommy Thompson and Becky Lavender. A generative grammar approach for action-adventure map generation in the legend of zelda. In *12th International Conference on the Foundations of Digital Games (FDG)*, 2017.

206. David Thue and Vadim Bulitko. Procedural game adaptation: Framing experience management as changing an mdp. In *8th AAAI Conference on Artificial Intelligence and Interactive Digital Entertainment (AIIDE)*, 2012.

207. Julian Togelius, Mike Preuss, Nicola Beume, Simon Wessing, Johan Hagelbäck, Georgios N Yannakakis, and Corrado Grappiolo. Controllable procedural map generation via multiobjective evolution. *Genetic Programming and Evolvable Machines*, 14(2):245–277, 2013.

208. Julian Togelius and Jurgen Schmidhuber. An experiment in automatic game design. In *2008 IEEE Symposium On Computational Intelligence and Games*, pages 111–118. Citeseer, 2008.

209. Julian Togelius, Georgios N Yannakakis, Kenneth O Stanley, and Cameron Browne. Search-based procedural content generation: A taxonomy and survey. *IEEE Transactions on Computational Intelligence and AI in Games*, 3(3):172–186, 2011.

210. Ruben Rodriguez Torrado, Ahmed Khalifa, Michael Cerny Green, Niels Justesen, Sebastian Risi, and Julian Togelius. Bootstrapping conditional gans for video game level generation. In *2nd IEEE Conference on Games (CoG)*, pages 41–48. IEEE, 2020.

211. Mike Treanor, Bryan Blackford, Michael Mateas, and Ian Bogost. Game-o-matic: Generating videogames that represent ideas. *3rd Workshop on Procedural Content Generation (PCG)*, pages 1–8, 2012.

212. Mike Treanor, Alexander Zook, Mirjam P Eladhari, Julian Togelius, Gillian Smith, Michael Cook, Tommy Thompson, Brian Magerko, John Levine, and Adam Smith. Ai-based game design patterns. In *10th International Conference on the Foundations of Digital Games (FDG)*, 2015.

213. Yudai Tsujino and Ryosuke Yamanishi. Dance dance gradation: a generation of fine-tuned dance charts. In *17th Conference on Entertainment Computing*, pages 175–187. Springer, 2018.

214. Jaap van Muijden. Gpu-based run-time procedural placement in horizon: Zero dawn. 2019 Game Developer's Conference (GDC), 2019.

215. Don Van Ravenzwaaij, Pete Cassey, and Scott D Brown. A simple introduction to markov chain monte–carlo sampling. *Psychonomic bulletin & review*, 25(1):143–154, 2018.

216. Judith van Stegeren and Jakub Myśliwiec. Fine-tuning gpt-2 on annotated rpg quests for npc dialogue generation. In *16th International Conference on the Foundations of Digital Games (FDG) 2021*, pages 1–8, 2021.

217. Ashish Vaswani, Noam Shazeer, Niki Parmar, Jakob Uszkoreit, Llion Jones, Aidan N. Gomez, Lukasz Kaiser, and Illia Polosukhin. Attention is all you need. In Isabelle Guyon, Ulrike von Luxburg, Samy Bengio, Hanna M. Wallach, Rob Fergus, S. V. N. Vishwanathan, and Roman Garnett, editors, *Advances in Neural Information Processing Systems 30: Annual Conference on Neural Information Processing Systems 2017, December 4-9, 2017, Long Beach, CA, USA*, pages 5998–6008, 2017. URL https://proceedings.neurips.cc/paper/2017/hash/3f5ee243547dee91fbd053c1c4a845aa-Abstract.html.

218. Vanessa Volz, Niels Justesen, Sam Snodgrass, Sahar Asadi, Sami Purmonen, Christoffer Holmgård, Julian Togelius, and Sebastian Risi. Capturing local and global patterns in procedural content generation via machine learning. In *2nd IEEE Conference on Games (CoG)*, pages 399–406. IEEE, 2020.

219. Vanessa Volz, Jacob Schrum, Jialin Liu, Simon M Lucas, Adam Smith, and Sebastian Risi. Evolving mario levels in the latent space of a deep convolutional generative adversarial network. In *20th ACM Genetic and Evolutionary Computation Conference (GECCO)*, pages 221–228, 2018.

220. Nick Walton. Ai dungeon 2: creating infinitely generated text adventures with deep learning language models, 2020.

221. Max Woolf. Create your own Magic: The Gathering cards with AI! https://twitter.com/minimaxir/status/1148610470594539520, 2019. Accessed: 2022-06-10.

222. Yuhuai Wu, Elman Mansimov, Roger B. Grosse, Shun Liao, and Jimmy Ba. Scalable trust-region method for deep reinforcement learning using kronecker-factored approximation. In Isabelle Guyon, Ulrike von Luxburg, Samy Bengio, Hanna M. Wallach, Rob Fergus, S. V. N. Vishwanathan, and Roman Garnett, editors, *Advances in Neural Information Processing Systems 30: Annual Conference on Neural Information Processing Systems 2017, December 4-9, 2017, Long Beach, CA, USA*, pages 5279–5288, 2017. URL https://proceedings.neurips.cc/paper/2017/hash/361440528766bbaaaa1901845cf4152b-Abstract.html.

223. Adeel Zafar. *Procedural Content Generation for General Video Game Level Generation*. PhD thesis, National University of Computer and Emerging Sciences, 2020.

224. Adeel Zafar, Shahbaz Hassan, et al. Corpus for angry birds level generation. In *2nd International Conference on Computing, Mathematics and Engineering Technologies (iCoMET)*, pages 1–4. IEEE, 2019.

225. Yahia Zakaria, Magda Fayek, and Mayada Hadhoud. Procedural level generation for sokoban via deep learning: An experimental study. *IEEE Transactions on Games*, pages 1–1, 2022. https://doi.org/10.1109/TG.2022.3175795.

226. Song-Hai Zhang, Shao-Kui Zhang, Yuan Liang, and Peter Hall. A survey of 3d indoor scene synthesis. *Journal of Computer Science and Technology*, 34(3):594–608, 2019.

227. Yuanlin Zhang and Roland H. C. Yap. Making AC-3 an optimal algorithm. In Bernhard Nebel, editor, *17th International Joint Conference on Artificial Intelligence, IJCAI 2001, Seattle, Washington, USA, August 4-10, 2001*, pages 316–321. Morgan Kaufmann, 2001.